5th edition

geog.2

geography for key stage 3

<rosemarie gallagher>
<richard parish>

OXFORD
UNIVERSITY PRESS

Great Clarendon Street, Oxford, OX2 6DP, United Kingdom

Oxford University Press is a department of the University of Oxford. It furthers the University's objective of excellence in research, scholarship, and education by publishing worldwide. Oxford is a registered trade mark of Oxford University Press in the UK and in certain other countries

© RoseMarie Gallagher, Richard Parish 2020

The moral rights of the authors have been asserted

Database right of Oxford University Press (maker) 2020

First published in 2001
Second edition 2005
Third edition 2008
Fourth edition 2014
Fifth edition 2020

All rights reserved. No part of this publication may be reproduced, stored in a retrieval system, or transmitted, in any form or by any means, without the prior permission in writing of Oxford University Press, or as expressly permitted by law, by licence or under terms agreed with the appropriate reprographics rights organization. Enquiries concerning reproduction outside the scope of the above should be sent to the Rights Department, Oxford University Press, at the address above.

You must not circulate this work in any other form and you must impose this same condition on any acquirer

British Library Cataloguing in Publication Data
Data available

ISBN 978-0-19-848915-3

11

The manufacturing process conforms to the environmental regulations of the country of origin.

Printed and bound in Great Britain by Bell & Bain Ltd, Glasgow

Acknowledgements
The publisher and authors would like to thank the following for permission to use photographs and other copyright material:

Cover: OUP/Shutterstock. **p5:** Suzanne Long/Alamy; **p6(t):** Chronicle/Alamy ; **p6(m):** GL Archive/Alamy ; **p7(tr):** Cultura Creative (RF)/Alamy ; **p7(b):** O Farooque; **p8(t):** Field Studies Council; **p8(b):** Kate Stockings; **p9(tl to br):** Skyscan Photolibrary/Alamy; Hufton+Crow-VIEW/Alamy; British Retail Photography/Alamy; Arcaid Images/Alamy; Shutterstock; View Pictures/Getty; **p10(t):** Suzanne Long/Alamy; **p10(m):** Field Studies Council; **p10(b):** laboratory/Alamy; **p12(t):** R Gallagher; **p12(b):** O Farooque; **p14(t):** Shutterstock; **p14:** heatmap courtesy of David Morgan (Field Studies Council); **p14(b):** O Farooque; **p17:** Apollo Imagery; **p18:** imageBROKER/Alamy; **p19:** Shutterstock; **p20:** Imaginechina Limited/Alamy; **p22(l):** Shutterstock; **p22(m):** Bill Bachman/Alamy; **p22(r):** Shutterstock; **p23:** NASA; **p25(tl to br):** Shutterstock; Jake Lyell/Alamy; Colin Harris/era-images/Alamy; DCPhoto/Alamy; eddie linssen/Alamy; Eden Breitz/Alamy; **p26(l):** Bridgeman Images; **p26(m):** Granger/Shutterstock; **p26(r):** agefotostock/Alamy; **p28(t):** Richard Green/Alamy ; **p28(tl to br):** Rolex Dela Pena/EPA/Shutterstock; Ian Leonard/Alamy ; Photofusion/Getty; Shutterstock; ROBERT BROOK/SCIENCE PHOTO LIBRARY; peter jordan/Alamy ; **p29(l):** Keren Su/China Span/Alamy ; **p29(m):** Shutterstock; **p29(r):** Paulette Sinclair/Alamy; **p30:** Shutterstock; **p31:** Image navi - QxQ images/Alamy; **p32:** Celia McMahon/Alamy; **p33:** Peter Adams Photography Ltd/Alamy; **p35(l):** Hirotaka Ihara/Alamy; **p35(r):** Shutterstock; **p36:** FALKENSTEINFOTO/Alamy ; **p37:** Manchester Libraries, Information and Archives ; **p38:** lowefoto/Alamy; **p39(tl to br):** GRANT ROONEY PREMIUM/Alamy; AFP/Getty; Roy Conchie/Alamy; Shutterstock; lowefoto/Alamy; Christian Richters-VIEW/Alamy; Simon Stacpoole/Offside/Getty; Terry Waller/Alamy; **p40:** Shutterstock; **p42(t):** Shutterstock; **p42(m):** Shutterstock; **p42(b):** Jane Sweeney/Robert Harding ; **p43(t):** Joerg Boethling/Alamy ; **p43(b):** Shutterstock; **p44(t):** Shutterstock; **p44(b):** Shutterstock; **p46(t):** PIUS UTOMI EKPEI/Getty; **p46(b):** Frédéric Soltan/Getty; **p47(t):** Thomas Mukoya/Reuters; **p47(l):** PIUS UTOMI EKPEI/Getty; **p47(r):** Shutterstock; **p48:** Shutterstock; **p49(tl to br):** Martin Bond/Alamy; Shutterstock; Bloomberg/Getty; _ultraforma_/Getty; Shutterstock; Shutterstock; Alan Gallery/Alamy; Cavan Images/OFFSET by Shutterstock; Shutterstock; **p50:** A.P.S. (UK)/Alamy; **p51(tl):** Stephen Finn/Alamy ; **p51(tr):** Ernie Janes/Alamy ; **p51(bl):** Roger Driscoll/Alamy; **p51(br):** pcpexclusive/Alamy ; **p53(l):** www.michaelmarten.com; **p53(r):** www.michaelmarten.com; **p54:** PictureLake/iStockphoto; **p55:** david a eastley/Alamy ; **p57(t):** Colin Allen/123RF; **p57(b):** Adrian Warren/Last refuge; **p58(tl to br):** Keyworded/Alamy; Shutterstock; Hemera Technologies/PHOTOS.com/Getty; Peter Titmuss/Alamy ; ickos/iStockphoto; Graham Prentice/Alamy ; Ingolf Pompe 65/Alamy ; Stephen Hird/Reuters; Keith Bowser/Alamy ; **p60:** Robert Taylor/Alamy ; **p61(t):** Visit Cornwall; **p61:** © Crown copyright 2020 OS100000249; **p61(bl):** Shutterstock; **p61(br):** Apollo Imagery; **p62(t):** Phil Noble/Reuters; **p62(m):** FLPA/Alamy; **p62(bl):** Arch White/Alamy ; **p62(br):** Dobson Agency/Shutterstock; **p64(t):** John Worrall/Alamy; **p64(b):** Mike Page; **p65(l):** Albanpix/Shutterstock; **p65(r):** Mike Page; **p66:** Hull News & Pictures Ltd; **p67(t):** Andrew Holt/Photographers Choice/Getty; **p67(b):** Shutterstock; **p68(t):** imageBROKER/Alamy ; **p68(b):** Alan Curtis/Alamy ; **p69:** Shutterstock; **p70(l):** Andrew Fox/Alamy ; **p70(r):** CamNews/Alamy ; **p71:** Shutterstock; **p72:** Daniel Valla FRPS/Alamy; **p73:** Greenshoots Communications/Alamy ; **p74:** DBURKE/Alamy; **p75(tl):** Kevin Britland/Alamy ; **p75(tr):** Shutterstock; **p75(b):** Shutterstock; **p76:** Shutterstock; **p78:** courtesy of Dundee Satellite Receiving Station; **p79:** Simon Maycock/Alamy ; **p80:** Ejla/iStockphoto; **p81(t, m):** Corel/OUP; **p81(b):** Shutterstock; **p82(t):** LPhot Paul Halliwell/Ministry of Defence via Getty; **p82(bl):** NASA; **p82(br):** NOAA via Getty; **p84:** SEYLLOU/AFP/Getty; **p86:** Shutterstock; **p88:** KARIM AGABI/EURELIOS/SCIENCE PHOTO LIBRARY; **p89:** Alamy; **p90:** Guy Bell/Alamy ; **p91:** Christopher Furlong/Getty; **p92(t):** The Natural History Museum/Alamy; **p92(b):** Shutterstock; **p94(tl):** William Crawford, Integrated Ocean Drilling Program, U.S. Implementing Organization (IODP USIO); **p94(tr):** MAURO FERMARIELLO/SCIENCE PHOTO LIBRARY; **p94(bl):** BRITISH ANTARCTIC SURVEY/SCIENCE PHOTO LIBRARY; **p94(br):** imageBROKER/Alamy; **p95(b):** NASA; **p96(l):** NASA; **p96(r):** Shutterstock; **p97(tl to br):** keith morris news/Alamy; Ashley Cooper pics/Alamy; Anadolu Agency/Getty; Michael Routh/Alamy; Mike Goldwater/Alamy; Nigel Cattlin/Alamy; **p98:** dpa picture alliance/Alamy; **p100(tl to br):** Simon Tilley/Alamy; Phil Rees/Alamy; Peter Cripps/Alamy; NASA; Shutterstock; Mike Goldwater/Alamy; **p101:** Lexie Harrison-Cripps/Alamy; **p102(t):** Ashley Cooper/Science Photo Library; **p102(m):** Sakis Papadopoulos/Getty; **p102(b):** Ashley Coope/Getty; **p103:** Koolstock/Radius Images; **p104tl):** Shutterstock; **p104(tr):** Shutterstock; **p104(b):** ZUMA Press, Inc./Alamy; **p105, 118(t):** PLANETOBSERVER/SCIENCE PHOTO LIBRARY; **p109:** Xinhua/Alamy ; **p110(tl to br):** Shutterstock; Shutterstock; guenterguni/iStockphoto; Shutterstock; Shutterstock; Shutterstock; **p113(t):** Shutterstock; **p113(bl):** Shutterstock; **p113(br):** NASA; **p115(tl):** Arterra Picture Library/Alamy; **p115(tr):** Shutterstock; **p116(tl to br):** Dmitry Deshevykh/Alamy ; **p116:** Elena Shchipkova/123RF; **p116:** Shutterstock; **p116:** Shutterstock; **p116:** Shutterstock; **p116:** Dmytro Korolov/123RF; **p117(l):** Shutterstock; **p117(r):** Shutterstock; **p118(b):** Sue Flood/Alamy; **p119(tl to br):** Li Ding/Alamy; Cheryl Rinzler/Alamy; Jarmo Piironen/Alamy; View Stock/Alamy; Gabbro/Alamy; Richard Ellis/Alamy; **p120(t):** Shutterstock; **p120(b):** Topical Press Agency/Getty; **p121(tl):** Keystone-France/Getty; **p121(tm):** AFP/Getty; **p121(tr):** ZUMA Press, Inc./Alamy; **p121(b):** Imaginechina Limited/Alamy; **p123:** ITAR-TASS News Agency/Alamy; **p124(l):** Imaginechina Limited/Alamy; **p124(r):** Ed Brown/Alamy; **p125:** Timothy Allen/Getty; **p127(tl):** Kevin Foy/Alamy; **p127(tr):** Sean Pavone/Alamy; **p127(b):** Lou-Foto/Alamy; **p128(tl):** B. Hall/Getty; **p128(tr):** Shutterstock; **p128(b):** Qilai Shen/Getty; **p129(tl):** Shutterstock; **p129(tr):** Shutterstock; **p129(b):** Imaginechina Limited/Alamy; **p130(tl):** Zhang Peng/Getty; **p130(tr):** dbimages/Alamy; **p130(b),131(t):** Zhang Peng/Getty; **p131(b):** View Stock/Alamy; **p132(t):** Bloomberg/Getty; **p132(m):** Wang He/Getty; **p132(bl):** Kyodo News/Getty; **p132(br):** Zhang Peng/Getty; **p133(t):** Imaginechina Limited/Alamy; **p133(m):** Kevin Frayer/Getty; **p133(b):** Ian Teh/Panos Pictures; **p134(t):** Xinhua/Alamy; **p134(b):** JOHANNES EISELE/Getty; **p135(t):** AB Forces News Collection/Alamy; **p135(b):** Paula Bronstein/Getty;

Artwork by Kamae Design, Mike Phillips Steve Evans, Ian West, Giorgio Bacchin, NAF, Ruth Palmer. Page layout by Kamae Design.

Every effort has been made to contact copyright holders of material reproduced in this book. Any omissions will be rectified in subsequent printings if notice is given to the publisher.

The Ordnance Survey material on pp 61 and 138 is reproduced with the permission of the Controller of Her Majesty's Stationery Office © Crown Copyright. Ordnance Survey (OS) is the national mapping agency for Great Britain, and a world-leading geospatial data and technology organisation. As a reliable partner to government, business and citizens across Britain and the world, OS helps its customers in virtually all sectors improve quality of life.

The changes in this edition of geog.2 are the result of comments from many people. We would like to thank the teachers who came together in focus groups to discuss the course, and Kate Stockings who provided a thoughtful and constructive overview. Thanks to David Morgan of the Field Studies Council and Sylvia Knight of the Royal Meteorological Society, who reviewed material. A special thank you to Garaeth Davies.

Note that the content of any direct speech attributed to characters in this book is based on information from reliable sources. Regarding maps used in this textbook, OUP takes no stance in contested territorial claims.

The manufacturer's authorised representative in the EU for product safety is Oxford University Press España S.A. of El Parque Empresarial San Fernando de Henares, Avenida de Castilla, 2 – 28830 Madrid (www.oup.es/en or product.safety@oup.com). OUP España S.A. also acts as importer into Spain of products made by the manufacturer.

Contents

1 Fieldwork, and GIS — 5
- 1.1 The fieldwork that changed the world — 6
- 1.2 What kind of fieldwork will you do? — 8
- 1.3 What are the stages in fieldwork? — 10
- 1.4 A sample fieldwork report — 12
- 1.5 What is GIS? — 14
- 1.6 GIS in fighting crime — 16
- Fieldwork, and GIS Check — 18

2 Population — 19
- 2.1 How is Earth's population changing? — 20
- 2.2 So where is everyone? — 22
- 2.3 Population growth around the world — 24
- 2.4 How is the UK's population changing? — 26
- 2.5 What is our impact on our planet? — 28
- 2.6 What does the future hold? — 30
- Population Check — 32

3 Urbanisation — 33
- 3.1 How did our towns and cities grow? — 34
- 3.2 Manchester's story – part 1 — 36
- 3.3 Manchester's story – part 2 — 38
- 3.4 Urbanisation around the world — 40
- 3.5 Push and pull factors — 42
- 3.6 It's not all sunshine! — 44
- 3.7 Life in the slums — 46
- 3.8 How can we make cities more sustainable? — 48
- Urbanisation Check — 50

4 Coasts — 51
- 4.1 What causes waves and tides? — 52
- 4.2 What work do the waves do? — 54
- 4.3 Which landforms do the waves create? — 56
- 4.4 What do we use the coast for? — 58
- 4.5 Your holiday in Newquay — 60
- 4.6 Storm surge! — 62
- 4.7 How long can Happisburgh hang on? — 64
- 4.8 How can we protect places from the sea? — 66
- Coasts Check — 68

5 Weather and climate — 69
- 5.1 Weather: what, why, and where? — 70
- 5.2 How is heat carried around Earth? — 72
- 5.3 Air pressure and our weather — 74
- 5.4 Why is our weather so changeable? — 76
- 5.5 What's a depression? — 78
- 5.6 More about rain … and clouds — 80
- 5.7 What's a tropical cyclone? — 82
- 5.8 Climate and climate factors — 84
- 5.9 So what's the UK's climate like? — 86
- 5.10 Climates around the world — 88
- Weather and climate Check — 90

6 Climate change — 91
- 6.1 Earth's climate – always changing! — 92
- 6.2 The climate detectives — 94
- 6.3 How is Earth's climate changing today? — 96
- 6.4 This time … is it us? — 98
- 6.5 Local actions, global effects — 100
- 6.6 What can we do? — 102
- Climate change Check — 104

7 Asia — 105
- 7.1 What and where is Asia? — 106
- 7.2 Asia's countries and regions — 108
- 7.3 What's Asia like? — 110
- 7.4 What are Asia's main physical features? — 112
- 7.5 Asia's population — 114
- 7.6 Asia's biomes — 116
- Asia Check — 118

8 China — 119
- 8.1 China: an overview — 120
- 8.2 A little history — 122
- 8.3 Mainland China's physical geography — 124
- 8.4 Where is everyone? — 126
- 8.5 How Shenzhen became a megacity — 128
- 8.6 Life in rural China — 130
- 8.7 What about the environment? — 132
- 8.8 What's the Belt and Road Initiative? — 134
- China Check — 136

★ Command words: a summary — 137

Ordnance Survey symbols — 138
Map of the British Isles — 139
Map of the world — 140
Glossary — 142
Index — 144

Really?

1 Fieldwork, and GIS

Me?

1.1 The fieldwork that changed the world

 Read about Doctor John Snow, whose fieldwork has helped to save millions of people from death by cholera.

Help! Cholera!

On 31 August 1854, a cholera outbreak hit the area called Soho, in London. Within three days, 127 people had died. Within ten days, 500 were dead. Many terrified people closed up their homes and fled.

At that time nobody knew what caused cholera, or how to treat it. Some thought you got it by breathing 'bad' air from things that were rotting. (Today we know it is caused by bacteria, and is quite simple to treat.)

Doctor Snow on the case

John Snow was a doctor in London at the time. He had an idea or **hypothesis** that cholera was spread by water, not air. He decided to investigate.

So he turned into a geographer! He started with a map of the Soho area, and marked on it all the households where people had died. He also marked the water pumps. And then he looked for patterns.

He noticed a cluster of deaths around a pump in Broad Street. But there were deaths close to other pumps too. On asking around, he found that even those victims had used the Broad Street pump. Some were children who had passed it on their way to school.

The pump is shut down

Doctor Snow convinced the council that the water from the Broad Street pump was the source of the cholera. So the pump handle was removed. (But by then the outbreak was already in decline, since so many people had fled.)

How did the water get infected?

In those days, most people in London got their water from street pumps. (Only the wealthier had piped water.)

And there was no sewage system like today. Toilet waste fell into a smelly pit under your house, called a **cesspool**. When it was full, you paid people to empty it.

The cholera outbreak had begun with a baby girl in Broad Street. She had been infected from elsewhere. And her mother had dropped the baby's waste, full of cholera bacteria, into the cesspool.

But the cesspool was leaky. So the bacteria made their way into the well under the Broad Street pump.

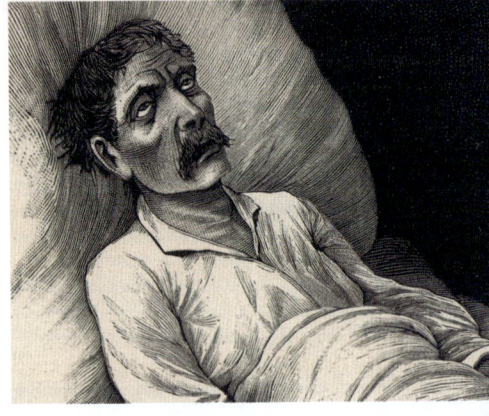

▲ *A cholera victim in the 19th century. He has severe diarrhoea and vomiting. He will die from dehydration.*

▲ *John Snow, 1813 – 1858. A doctor – and a geographer by accident!*

▲ *How the water at the Broad Street pump became infected.*

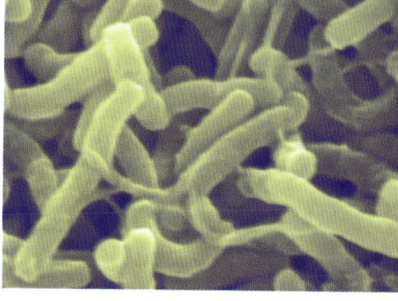

▲ Cholera bacteria, magnified. Doctor Snow did not have a microscope powerful enough to see them. Boiling infected water will kill them.

Key
- deaths from cholera (the more deaths, the taller the bar)
- public water pumps

◂ Doctor Snow's map of the Soho area, where cholera struck.

London gets cleaned up

Doctor Snow's work helped to convince people that London must get rid of cesspools, and build a proper sewage system, with flush toilets. Work began on the sewage system in 1859. It is being updated today.

It was fieldwork

Doctor Snow's work is an example of **fieldwork**. In fieldwork you collect information or **data** in order to test a hypothesis, or answer a question.

Doctor Snow's fieldwork was a big step in proving that cholera is spread by infected water. (The bacteria can also get into food, from infected water.) So his work has helped to save millions of lives around the world.

You will do fieldwork too. It won't be as dramatic as Doctor Snow's – but you will learn about the process. And who knows where that will lead!

▲ Broad Street is now called Broadwick Street. This pump with no handle is a memorial to Dr Snow.

Your turn

1. What caused the cholera outbreak in Soho in 1854?
2. Define these terms. (Glossary?)
   a hypothesis   b fieldwork   c data
3. State the hypothesis for Dr Snow's fieldwork, in five words.
4. State what information these give us, on Dr Snow's map:
   a the words   b the red bars   c the blue circles
5. From his map, why did Dr Snow deduce that the Broad Street pump was the source of the cholera?
6. Explain why it was essential for Dr Snow to:
   a use a map for his fieldwork
   b mark the pumps and deaths in the correct places on it
7. During his fieldwork, Dr Snow interviewed people in Soho who had *not* caught cholera. Suggest two key questions he could have asked them, to help test his hypothesis.
8. To what extent did Dr Snow prove his hypothesis?
9. You are a public health official.
   There is a severe outbreak of food poisoning in your area. Your question is: *What is the source of this food poisoning?*
   a Outline what you could do to find the answer. Would a map help?
   b Suggest reasons why finding the answer may be more difficult today than in Dr Snow's time.

7

1.2 What kind of fieldwork will you do?

 In the last unit you read about Doctor Snow's fieldwork. Here we look at the kinds of fieldwork you might do.

Fieldwork in geography

You are a geography explorer. You usually explore from your desk. But sometimes you go out and explore in a real place. It's your **geography fieldwork**.

Your starting point

All fieldwork starts with …

- an **enquiry question**. For example Doctor Snow could have asked: *What is the cause of the cholera outbreak in Soho?*
- or a **hypothesis**, like the one Doctor Snow began with. A hypothesis is a statement you think is true – but you're not sure. His hypothesis was: *Cholera is spread in drinking water.*

It's easy to turn a hypothesis into an enquiry question, and vice versa.

The aim of fieldwork is to answer the enquiry question, or test the hypothesis to see if it's true.

Collecting data

To answer an enquiry question, or test a hypothesis, you collect **data**.

- This often means *counting* or *measuring* things. For example count shops. The data you obtain by counting and measuring is called **quantitative** data. (*Quantitative* data tells you *quantity*.)
- All other data is **qualitative** data. It's often about opinions and feelings. For example you might ask people how they feel about climate change. Photos and field sketches count as qualitative data.
- The data you and your class go out and collect is called **primary** data.
- You may find useful data on the internet too, or in books – or your teacher may give you some. This is called **secondary** data.

Look at the examples of fieldwork topics on the next page. Then try *Your turn*.

▲ *Fieldwork: measuring the width of a river's channel. Quantitative data!*

▲ *Fieldwork: giving scores for the quality of an environment. Qualitative data!*

Your turn

1. **a** From page 9, pick out:
   i a hypothesis ii an enquiry question
   b i Rewrite your hypothesis in **a** as an enquiry question.
   ii Rewrite your enquiry question in **a** as a hypothesis.

2. Pick out a topic from page 9, where you are likely to collect:
   a qualitative data   **b** quantitative data
   Explain each choice, to show that you understand what each term above means.

3. From page 9, choose one topic for:
   a a human enquiry   **b** a physical enquiry

4. Is your data primary data, or secondary, in these examples?
   a You measure the width of the river in different places.
   b You use an OS map to help you choose where to measure.

5. You can take photos and draw field sketches, during fieldwork.
   a Define the term *field sketch*.
   b Imagine you took the photo of the cliffs and beach on page 9. Now draw a labelled field sketch of the same scene. No need to show people! (Those barriers are gr_____ .)

6. Work with a partner to think up *three* new enquiry questions for geography fieldwork for your class. None from page 9!

Examples of enquiry questions and hypotheses

In human geography

Human geography is about places and structures we humans created, and how we live and work. Look at these 'human' fieldwork topics:

People are satisfied with the bus services in this area.

To what extent does the park serve the needs of wheelchair users?

Litter is a problem in the school grounds.

How far do the students in my year travel to school?

How far do people travel to shop here?

This shopping centre has had a negative impact on local shops.

In physical geography

Physical geography is about natural features and processes. Like rivers, coasts, weather. Some 'physical' fieldwork may overlap with human geography. Look:

What evidence is there of different coastal processes at work here?

Erosion is changing the coastline in this place.

The width of the river increases downstream.

What evidence is there of different processes at work along the river?

The temperature varies around the school grounds.

Where is the best place for a school picnic area?

1.3 What are the stages in fieldwork?

There are lots of different fieldwork topics … but the stages are the same for them all. Find out more here.

The stages in fieldwork

To make fieldwork easy for you, it is divided into clear stages. Look:

1. **Decide on your enquiry question or hypothesis.**

2. **Plan your fieldwork carefully.**

   What type of data do you need, to answer your enquiry question or test your hypothesis?
   How and where will you collect it, and how much do you need?
   List everything you will need to bring with you, and get it ready. This includes questionnaires or tables for recording data.
   How will you make sure you stay safe, and avoid accidents?

3. **Collect the data.**

   Count and measure accurately.
   Record carefully.
   Stay safe.

4. **Process and present the data, to make it easy to analyse.**

   For example do calculations, draw graphs, draw bar charts, annotate maps.

5. **Analyse the data.**

   What patterns and trends does it show?
   Does anything not fit the pattern? Can you suggest a reason?

6. **Draw conclusions from the data.**

   What is the answer to your enquiry question? … *or*
   Was your hypothesis correct?

7. **Evaluate your enquiry**

   Could you have done anything better? For example taken more measurements, or asked more people, or used a different method?
   Do you think your conclusion is reliable? Why do you think that?

At the end, you will write up the enquiry as a **report**. See Unit 1.4.

What if …
… your class did fieldwork on the moon?

What if …
… nobody did fieldwork?

▲ Stages 1 – 7 are the same for all fieldwork. Even if it's about the rainforest, or coral reefs as on page 5.

▲ Fieldwork to measure how fast water infiltrates soil. The faster it goes, the lower the flood risk when there's heavy rain.

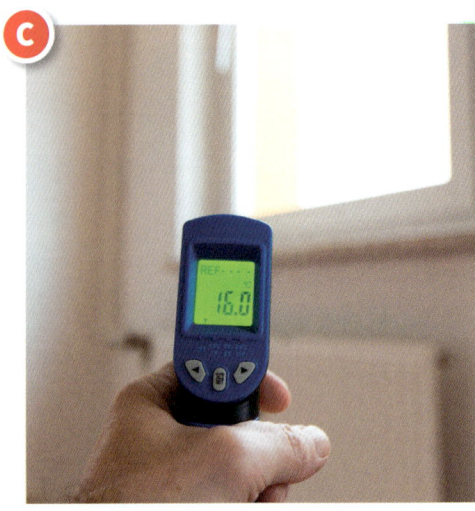
▲ Fieldwork to detect heat loss. Point the infrared thermometer at a surface to find its temperature. If heat is leaking out, that area will be cooler than the surrounding areas.

Your turn

Fieldwork, and GIS

D Some methods of collecting data

Ask people questions.
Measure things.
Count things.
Look around you, and write down what you see.
Smell the air.
Listen to the noise around you.
Take photos.
Draw a field sketch (a sketch of what you see).

E Two enquiry questions

1 Is litter a problem in the school grounds?
2 How do different surfaces affect the rate of infiltration – and therefore the flood risk?

F Items you might need for your fieldwork

a questionnaire, with the questions you will ask people
a table or tables to record data in
a pen or pencil (and an eraser might help)
a tally chart or a click meter, for keeping count
a measuring tape
a stopwatch or timer (could be on a mobile)
a thermometer
an anemometer (for measuring wind speed)
wellingtons
rubber gloves
high-visibility vest
a camera (could be on a mobile)
a mobile with an app to measure the loudness of noise
a notebook and/or clipboard
a map or plan of the fieldwork area

1 Look at the fieldwork stages on page 10.
  You carry out most of these in the classroom.
  a Which stage is usually carried out *outside* the classroom?
  b It's important to carry out stage 2 carefully. Explain why.
  c Look at the fieldwork topics on page 9. Choose one topic where you may need to take special care in stage 3, to keep yourself safe. Explain your choice.
  d In which stages might you use some maths?
  e Suggest a reason why stage 7 is important.

2 Look at enquiry E1 above. In which does it belong?
  a physical geography   b human geography

3 You are planning your fieldwork for E1. You are deciding what data you will need. Collecting data you won't use is a waste of time. From the list below, choose *five* sets of data that you will collect for E1. For each, explain your choice.
  a the places where you find litter (to be marked on a map)
  b the wind speed where litter is found
  c the type of litter it is
  d the number of pieces of litter at each place
  e the locations of any litter bins in the grounds
  f the condition of any litter bins (full? empty?)
  g photos of the litter / litter bins
  h any notices about litter around the school
  i the number of students in each class

4 From the data sets you chose in question 3, identify those which are:
  i quantitative (you count or measure)
  ii qualitative (may be a matter of opinion or choice)
  You can write down their letters (**a – i**) as your answer.

5 Look at the items in **F**. Select those you think you'll use in your fieldwork for **E1**. For each, explain your choice.

6 Now look at **E2**. The slower the *rate of infiltration*, the greater the chance of flooding when there is heavy rain.
  a Define the term in italics above. (Glossary?)
  b Photo **B** shows stage 3 of the fieldwork for **E2**.
    One student is pouring a measured amount of water onto soil, through a concrete cylinder.
    Identify four items the other student is using for this fieldwork. (Choose from **F**.)
  c Would you expect the students to get different results in different weather? Explain your answer.

7 a From page 9, choose another topic not mentioned here.
  b Select the method or methods of data collection that you'd expect to use for this topic, from **D**.
  c Now select the items you will use, from **F**. (You can add others if you need to.) For each, explain your choice.

8 We can help to limit climate change by making sure buildings are well insulated. So they don't leak heat – and we use less central heating fuel.
  a Define *insulated*. Glossary?
  b Decide on an enquiry question or hypothesis about the heat insulation in your school.
  c Outline how you and your class will carry out the fieldwork, using the special thermometer shown in **C**.

11

1.4 A sample fieldwork report

The fieldwork report below may help you to write your own. But it has gaps – and you have to complete them in *Your turn*!

GEOGRAPHY FIELDWORK REPORT   NAME: SAM GREY

ENQUIRY QUESTION: Is litter a problem in the school grounds?

Introduction
Litter makes a place look messy. It gives a bad impression. It can carry germs, and attract flies and rodents. It can harm birds and other wildlife too.

The aim of our fieldwork was to find out whether litter is a problem in the school grounds.

Planning
First the class discussed how to group the litter we might find, and what kind of table we needed to record it. When we had decided, Miss Benson got copies of the table printed out.

Miss Benson divided us into six teams. Some teams were bigger because they had a bigger area to cover. She gave each team a map of the school grounds, with the teams' areas marked on. This is secondary data.

I was in team 1, for area 1 on the map. I was in charge of the map for my team. Each team member had:

– a copy of the table to record the data

– a pen or pencil

– a clipboard.

Collecting the data
My team walked around our area. When we found litter we counted each type, and recorded the numbers in the table using tally marks. We also checked the litter bins, estimated how full they were, and I shaded in their icons on our map.

Processing and presenting the data
The teams combined their data into a master table (Figure 1). Then we created a master map, and and a key (Figure 2). This took quite a lot of time.

Analysing the data
Figure 2 shows the areas where the teams found litter.

– Most litter was found around the outdoor picnic area. It is near the shop in the dining hall, where you can buy snacks and sweets.

– The second biggest amount was found _____.
   This is probably because _____

– The third biggest amount was _____
   This is probably because _____

– The main type of litter was _____

– I also notice that _____

– Some litter was right beside litter bins, even though the bins were _____

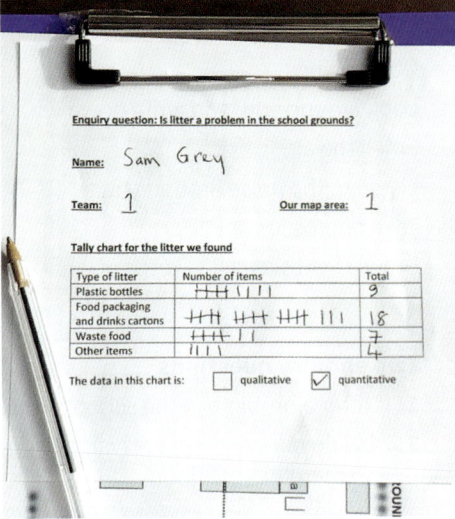

▲ Sam's completed tally chart.

Did you know?
Breaking down in the environment can take:
- 2 years for a banana skin
- 10 years for a plastic bag
- thousands of years for a plastic bottle.

▲ You'd never do this!

Did you know?
- It costs local councils in the UK about £1.50 to remove a single patch of chewing gum from a pavement.

Figure 1. The litter we found in the school grounds.

Areas where we found litter (see map)	Plastic bottles	Food packaging and drinks cartons	Waste food	Other items
1	9	18	7	4
2	4	14	2	2
3	3	8	4	3
4	2	10	0	3
5	0	5	1	1
6	14	13	5	4

Good work Sam! Just one thing – the results are not easy to compare, on the map. Find a more visual way to display the data!

Figure 2. The litter and the litter bins in each survey area.

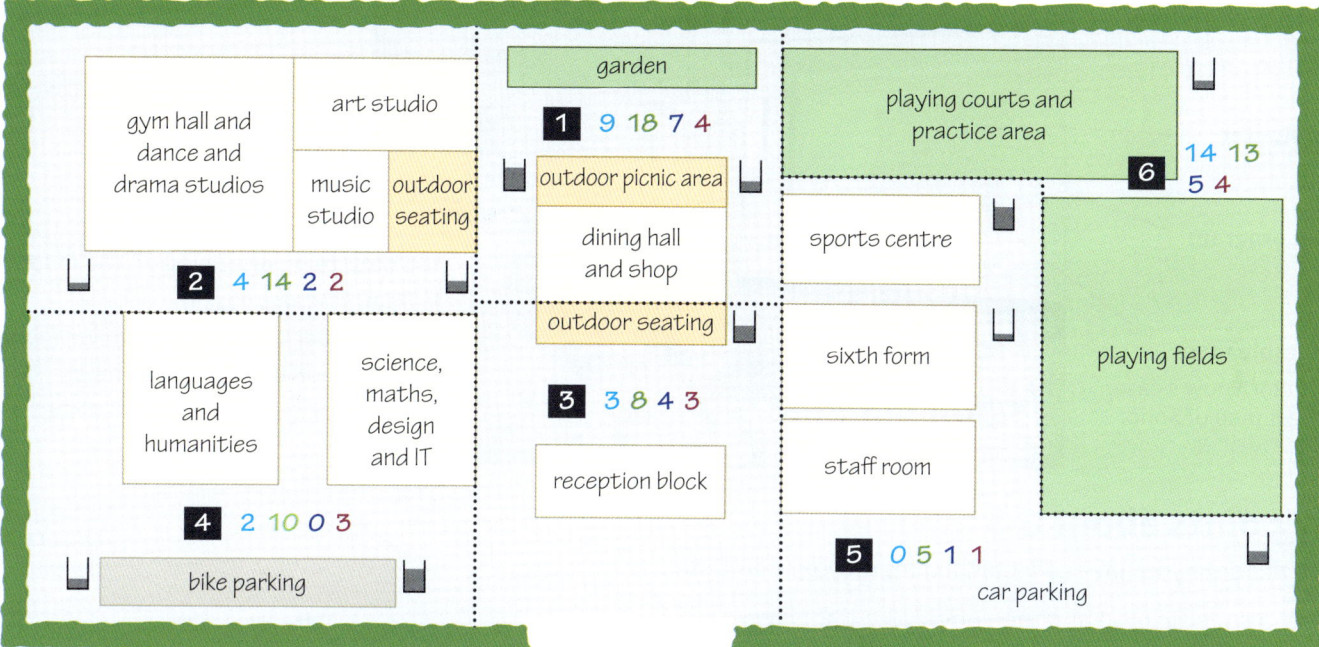

Conclusions

Our fieldwork showed that litter is _____

Especially worrying was the _____ because _____ .

Here are my suggestions for tackling the litter problem:

_____

Evaluation

In my opinion we did a good job with our fieldwork. It was a windy day so some litter may have blown away, but this does not affect my overall conclusion.
If we had litter pickers and sacks we could have picked up the litter too.

Key

1	team
······	border of team area
⌐⌐	litter bin and contents

Types of litter

5 plastic bottles
5 food packaging and drinks cartons
5 waste food
5 other items

Your turn

1. As you can see, Sam's report is not complete. That's up to you! So first, read it carefully. Study the map and key.

2. Sam's *analysis* is not complete. Copy it into your exercise book, but fill in each gap as you go. All the data you need is given above. Write as much as you wish for each gap.

3. Decide whether Sam's fieldwork has answered the enquiry question. Then copy and complete the *conclusions*. Write as much as you wish. Try to come up with good suggestions.

4. Think about Sam's *evaluation*. Would you change anything about the planning and data collection? If yes, explain what, and why.

5. Look at Miss Benson's comments about the data display.
   a. How would *you* improve the data display? (Use bar charts? divided bars? pie charts? scattergrams?)
   b. Using the method you think best, display the data for team 1 on a larger copy of area 1 of the map.

1.5 What is GIS?

 GIS lets you display and analyse fieldwork data on a map on a computer. Find out more here.

> **Did you know?**
> - Rainforest dwellers are learning to use GIS!
> - They map the trees, to help protect the forest from illegal logging.

What is GIS?

In Unit 1.1 you read about Doctor Snow. He marked data on his map by hand.

If Doctor Snow were alive today, he'd use **GIS** instead. GIS stands for **Geographic Information System**. It lets you display data on maps on a computer, and helps you to analyse it.

GIS has these components:

① **a computer**

② **a GIS program**

③ **a suitable map**
In Doctor Snow's case, a street map of Soho.

④ **the data to be displayed on the map**
In Doctor Snow's case the data is:
– the number of deaths at each location
– the location of the water pumps
Date can include photos and sketches.

The key points about GIS

- You save the data you have collected in files on your computer.
- You select a suitable map and then open your data files.
- Data must show up in the right places on the map or else it's no use. So each piece of data is tagged (linked) with the **latitude** and **longitude** of the place where it was collected.
- Each type of data is in a different layer on the map. So for Doctor Snow's map, one layer will show where the water pumps are and another will show the deaths. *And you can switch layers on and off.*

If you are Doctor Snow …

Imagine you are Doctor Snow. You have a smartphone, and GIS. This is how you might do your fieldwork. (You'll fill in a table like **A**.)

- Go to each house in each street in Soho. Knock on the door and ask:
  - Did anyone in this house die from cholera? If so, how many people?
  - Which pump do you get your water from?

  If nobody answers, the neighbours might know. Keep asking! (Knocking on strangers' doors is okay for Doctor Snow, but not students!)

- Fill in the data in your table, for each house where there was cholera.
- Use your mobile to find the latitude and longitude of the house, and add this data to your table too.
- Go to each local water pump, and record its latitude and longitude in a separate table. You could take photos too if that was useful.

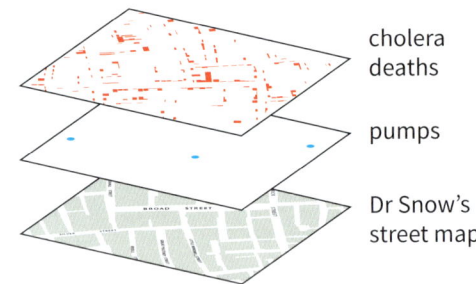

▲ Dr Snow's GIS map has two layers of data on top of the map. It's just like placing two see-through overlays on the map. And you can switch a layer on and off. Easy!

- cholera deaths
- pumps
- Dr Snow's street map

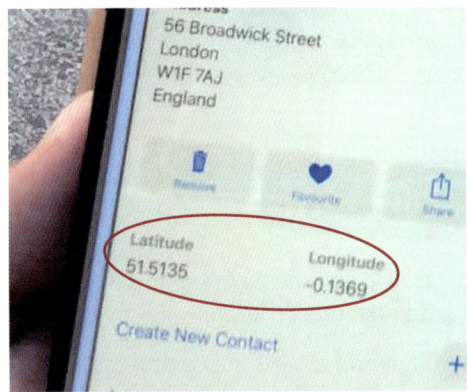

▲ Smartphones have built-in GPS. So they give latitude and longitude, in decimal form. The culprit pump in Soho was at 51.5135 °N, 0.1369 °W. (The – means 'west'.)

Fieldwork, and GIS

Doctor Snow's data

A shows the kind of table Doctor Snow could use to record deaths. (The pumps have their own table.)

Latitude and longitude are in decimal form. The minus signs for longitude mean *'west of the Prime Meridan (0°)'*.

A

Address	Latitude	Longitude	Number of deaths at address	Water pump used by residents
22 Berwick Street	51.5138	−0.1351	3	Broad Street
83 Berwick Street	51.5139	−0.1353	7	Broad Street
85 Berwick Street	51.5140	−0.1353	6	Broad Street

Deaths could be shown as bars on the 'deaths' layer, at the correct latitude and longitude on the map. The more deaths at an address, the taller the bar. (Look back at the bars on the map on page 7.)

The pumps will be on a second layer.

Doctor Snow's GIS map

B shows an up-to-date GIS map for Doctor Snow's data. (Broad Street is now called Broadwick Street.)

But here we don't show bars for deaths, unlike on page 7.

Instead we created a **heatmap**. It shows the density of deaths in the area. Look at the key. The brighter the yellow, the more deaths.

The heatmap shows clearly that the deaths are centred on one pump – the pump in Broadwick Street.

GIS gives you lots of choices for displaying data. It's brilliant!

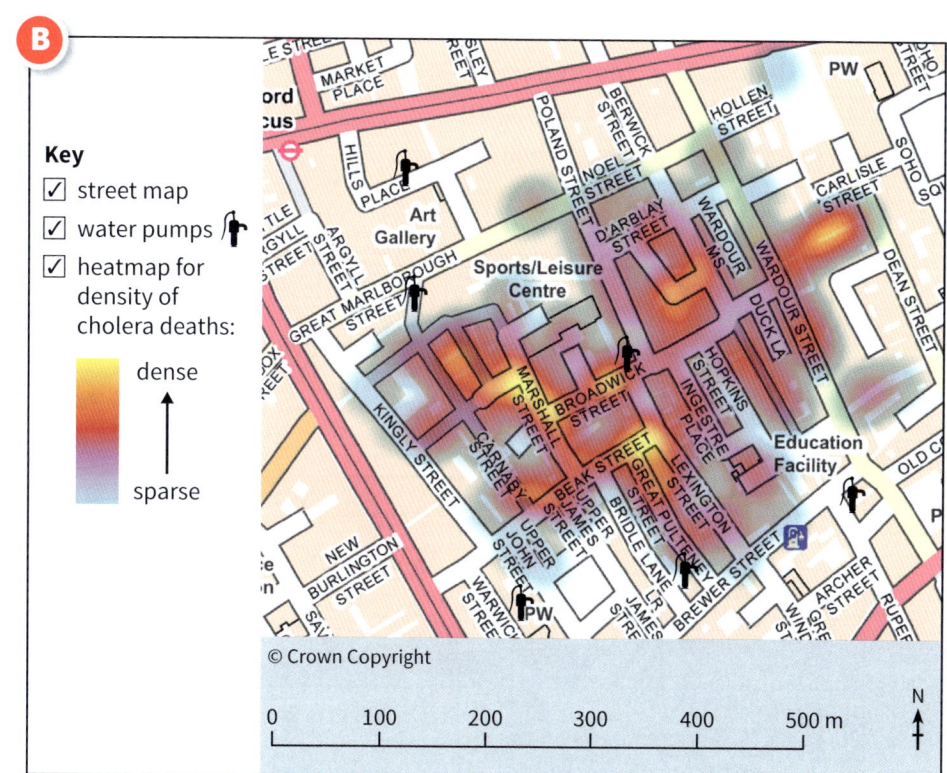

Your turn

1. **a** What does *GIS* stand for?
   b Name the four things you need, to use GIS.
2. Define these terms: **a** latitude **b** longitude
3. Explain *why* you need to record latitude and longitude with the data you collect, when using GIS.
4. Describe how Doctor Snow would have carried out his data collection if he'd had a smartphone and GIS.
5. State the two main differences between a GIS map and a paper map.
6. Look at map **B**. Do you think the heatmap is useful? Decide, and give your reasons.
7. Suppose you are looking at map **B** on a computer screen. What will happen if you untick the 'pumps' tick box?
8. Doctor Snow wants to rule out the pie shops in Soho as a source of cholera. So he will add a new layer called 'pie shops' to **B**, to show where they are.
   a Which layer would you turn off if you wanted to look for a connection between cholera deaths and pie shops?
   b Using **a** as an example, explain why it's useful to be able to switch layers on and off.
9. *GIS is more useful than a paper map, for presenting and analysing data.*
   a To what extent do you agree with the statement in italics?
   b When might it be better to use a paper map? Discuss!

15

1.6 GIS in fighting crime

The police do lots of fieldwork – and use GIS. Here you can explore a GIS crime map, and compare it with an aerial photo.

GIS for the police

The police are always out doing fieldwork!

They investigate crimes. They record the crimes and their locations. Then they bring the data into GIS. They analyse the maps to identify patterns and crime hotspots. So they can decide where more patrols are needed, for example.

Put your police hat on!

You are in charge of crime control for the area on the GIS map below. The coloured dots show where crimes occurred over the last six months.

The matching photo on page 17 will help you answer the *Your turn* questions.

Evening all!

We're doing our fieldwork.

Key for street map
- railway and railway bridges
- embankment

Businesses
- shops (all types)
- financial centres (banks, building societies, post office)
- places of entertainment (pubs, clubs, cafes, restaurants)

Abbreviations
- PW place of worship
- PO post office
- Mkr market
- Sta station

GIS layers
- ☑ street map
- ☐ aerial photo

Crime over last 6 months
- ☑ ● household burglary
- ☑ ● repeat household burglary
- ☑ ◆ break-ins to businesses
- ☑ ● assault (fighting)
- ☑ ■ theft of, or from, cars
- ☑ ● illegal dumping of rubbish
- ☑ ▲ vandalism
- ☑ ■ mugging

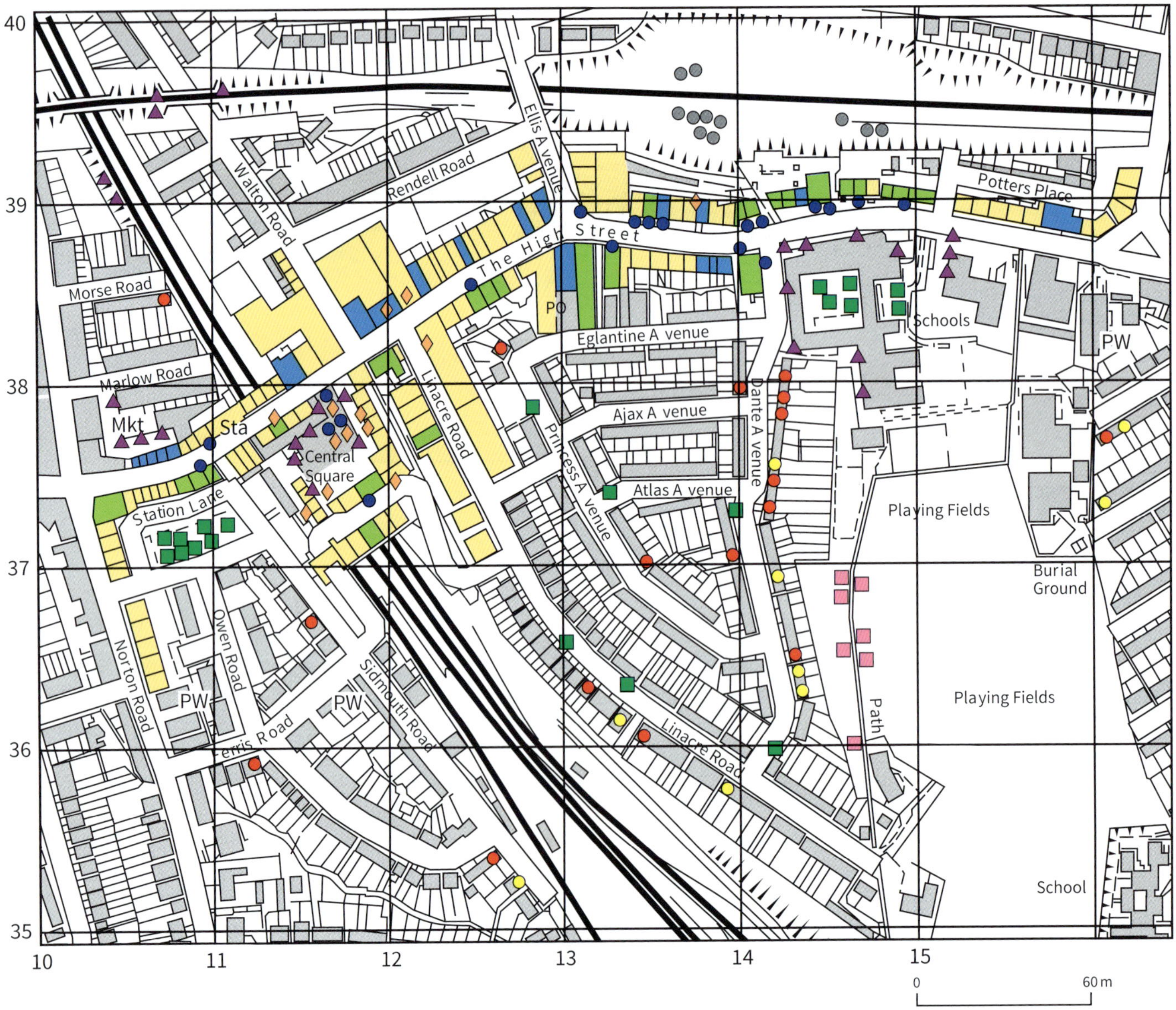

Fieldwork, and GIS

Your turn

1. The map shows there were several fights along one part of the High Street, in the last six months.
   a. Suggest a reason for this. (Check building use?)
   b. What could you do to prevent trouble there? See how many suggestions you can come up with.

2. Now look at square 1436.
   a. Identify the main crime here.
   b. Suggest a reason for this. (Check the aerial photo.)
   c. What could you do to reduce or prevent this crime here? Put your suggestions in order, best one first.

3. Yesterday two of your police team visited each house on the right of Dante Avenue (going north). They handed out special pens with invisible ink. People can use these to write their postcodes on valuable items such as computers.
   a. What is a *postcode*? Give an example.
   b. Why do the police want people to write their postcodes on things?
   c. Why did they choose that road?

4. Houses on the left of Dante Avenue are burgled far less often than those on the right. Suggest a reason.

5. Which grid square was worst for this crime? And why? (Check the photo.)
   a. theft of, or from, cars
   b. illegal dumping of rubbish

6. Vandalism is a problem too. Windows get broken, phone booths smashed, and walls sprayed with graffiti. There's a lot of it in squares 1438 and 1137. Suggest reasons for this.

7. Look at the aerial photo again.
   a. Is it a good match for the map? Give it a score out of 10.
   b. Should the police get rid of the map, and show the crime data on the photo instead? Explain your answer.
   c. In fact the police usually add an aerial photo as a layer in their GIS menus. Is this a good idea? Explain.

8. People take steps to deter criminals. You've been asked to identify the steps they take in your local area. Describe what primary data you plan to collect and record, using only your smartphone.

GIS makes me nervous.

1 Fieldwork, and GIS

How much have you learned about fieldwork, and GIS? Let's see.

A

1 a Write down the stages in geography fieldwork, in the correct order – without looking back in the chapter!
   b Then check if you got them right. If wrong, correct them.

2 Let's see if those fieldwork stages work for other tasks too. You have been given money to buy something you really want – for example a mobile or an electric scooter. But you don't know which brand to buy.
   a Write yourself an enquiry question for this dilemma.
   b Which of the other fieldwork stages might help you answer your enquiry question? Tick them on your list for **1b**.
   c You will collect data to answer your question in **a**.
      i Will you need to collect primary data? If so, where will you go to collect it?
      ii Where will you look for secondary data?
   d Write down two (imaginary) examples of data about your purchase – one qualitative and one quantitative.
   e After buying what you want, will it be useful to do an evaluation? Explain your answer.

3 Look back at Sam's table on page 13 (Figure 1). Your task is to present his data in a table like this instead.

Type of litter	Number found
Plastic bottles	
Food packaging and drinks cartons	
Waste food	
Other items	
Total number of items	

   a Copy and complete the table above, using the data from Sam's table.
   b Now present the data in your table in graphical form. (You could use a bar chart, divided bar, pie chart, or pictogram.)
   c Does your answer for **b** make it any easier to suggest solutions to the litter problem? Explain.

4 a What does *GIS* stand for?
   b Write out this paragraph, with the blue words unjumbled.
      In GIS, the atad is shown on a pam on the computer rescen. Each type of taad is in a different realy on the map, so you can witshc it on and off when you are looking for snatterp.
   c Latitude and longitude must be recorded for each piece of data for GIS. Explain why.
   d Suggest a way to find your exact latitude and longitude.

5 a Could Sam have used GIS in his fieldwork? (Unit 1.4.) Explain your answer.
   b Might Sam's map be more helpful, using GIS? (Think about how he might show exactly where they found litter, for example.)

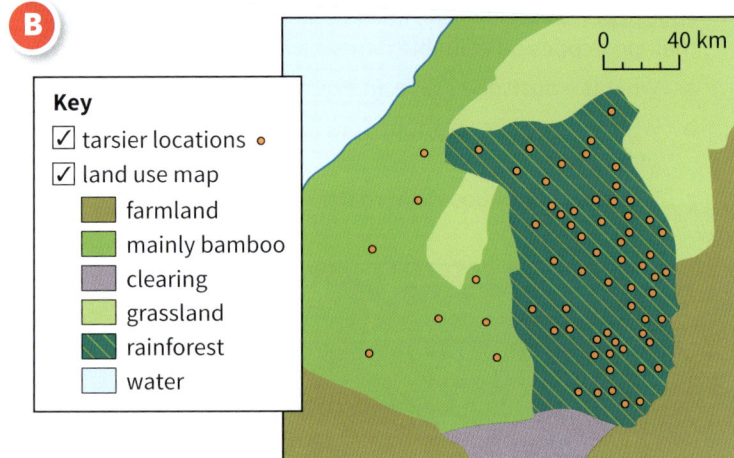

B

6 A shows a tarsier. These little animals once lived in rainforests everywhere. But now they are under threat. They are found only on a few islands in Asia. People are doing fieldwork on the tarsiers. They put tiny GPS collars on them, to track them. They use GIS to display the data.
   B shows the tarsiers in one area, at a moment in time.
   a What data would a *GPS collar* provide? (Glossary?)
   b Do tarsiers appear to live in isolation, or in groups? Check the scale, then justify your answer.
   c What clue(s) can you find that:
      i tarsiers do not like water?
      ii tarsiers live in trees?
   d Which do tarsiers seem to prefer: bamboo forest or rainforest?
   e People are burning down this rainforest, to get more farmland. Predict the impact of this on the tarsiers.
   f Could maps like B help people to protect the tarsiers? Explain.

7 You have met the seven stages of fieldwork. Which stage do you think is the most important? Justify your answer.

2 Population

2.1 How is Earth's population changing?

Here you'll discover how quickly our numbers are rising – and why.

Here we go!

Think about this… and be astonished!

10 000 years ago, there were about 5 million of us humans on Earth. By 1800 CE there were about 1 billion. By 2000, about 6 billion. And today, over 7.7 billion!

Did you know?
- Around 360 000 new babies are born every single day.

How does the population rise so fast?

Let's take a family in Britain as an example.

1750

In 1742, Bo and Ella fell in love. They got married and had 7 children. Sadly, 3 died at birth or soon after. The other 4 grew up and …

1780

… had children of their own. 2 of these died at birth, and 6 (around your age) died of smallpox. But **14** survived to be adults.

1820

12 of the 14 adults had children. 2 died at birth. 5 died in an outbreak of typhus, carried by lice. But **70** survived to be adults.

So by 1820, Bo and Ella had over 70 living descendants. By now they may have hundreds! So it is easy to see how the world's population rises fast.

But what about deaths?

Every year, millions of humans die. But the world population keeps rising each year, because there are more births than deaths.

For example in 2018, there were about 83 million more births than deaths around the world. So the population increased by about 83 million.

Life expectancy

Here is one big reason for population growth: people are living longer!

Life expectancy means how many years a new baby can expect to live for, *on average*. (Some might live to age 30, some 65, some 83.)

In 1742, when Bo met Ella, life expectancy for a new baby in Britain was under 40. This was mainly because of poor hygiene. Sewage and other waste was dumped in rivers, where people got their water. Infections spread easily.

A clean water supply, a sewage system, vaccines: all help to improve life expectancy. So more babies live to be adults, and have children of their own.

For a baby born in the UK today, life expectancy is 81, on average. (It is 83 for baby girls and 79 for baby boys. Females tend to live longer!)

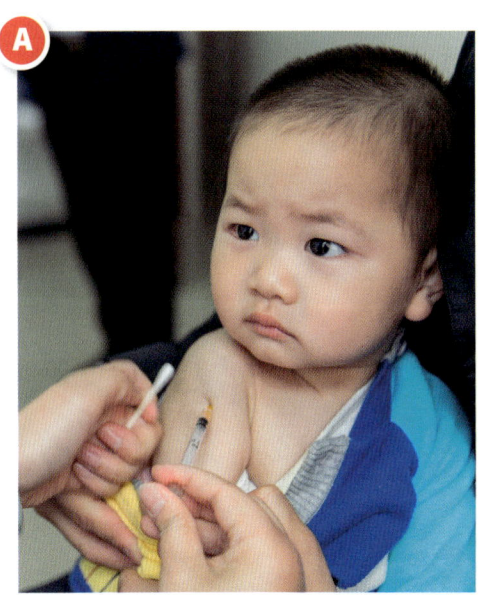

▲ *Why are you doing this to me?*

A graph of world population

Experts say that we humans (*Homo sapiens*) first appeared in Africa, over 300 000 years ago. We started migrating around Earth about 60 000 years ago.

Graph **B** shows how our population has grown since 10 000 BCE. Look how it shot up.

B

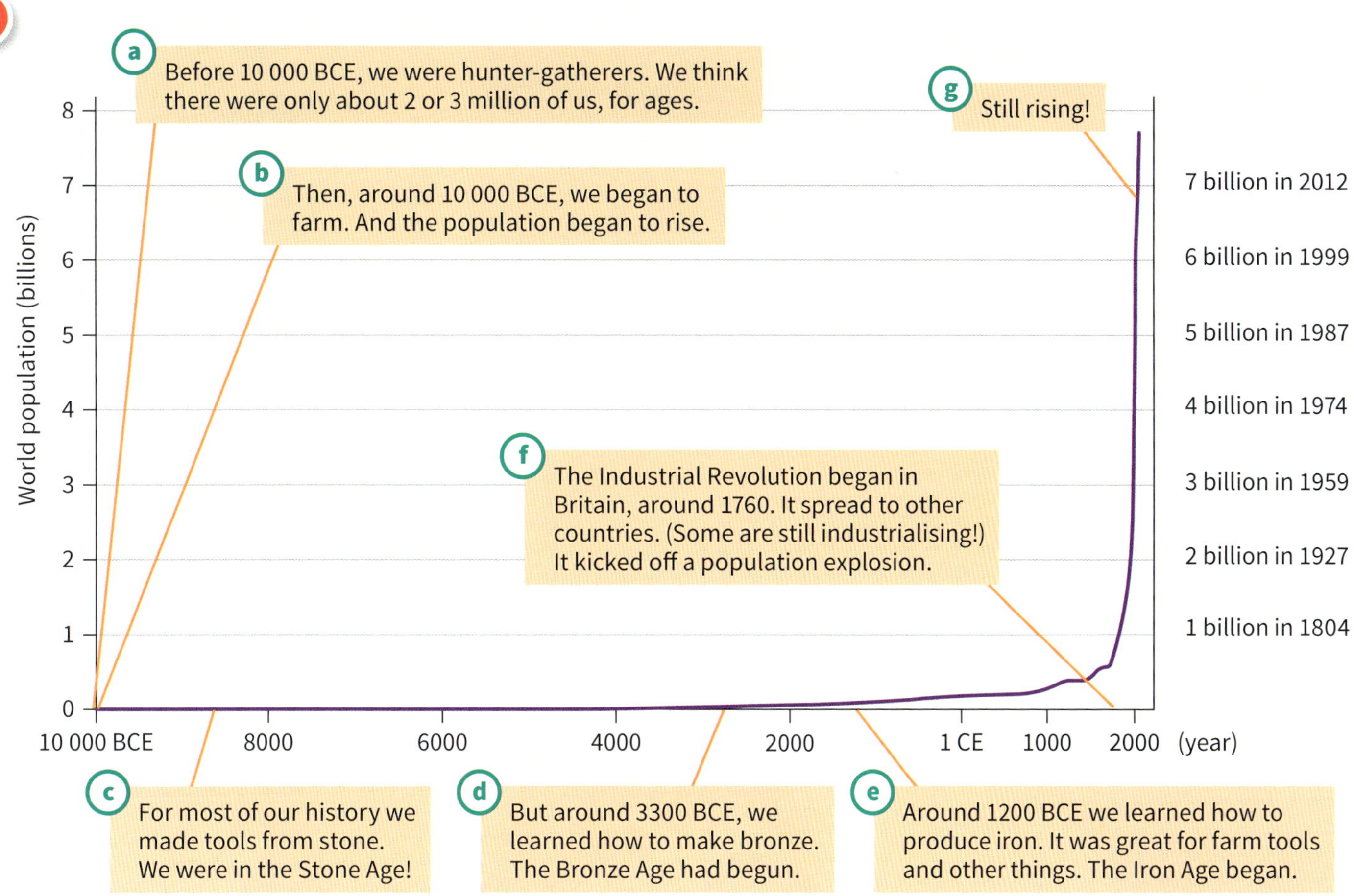
a Before 10 000 BCE, we were hunter-gatherers. We think there were only about 2 or 3 million of us, for ages.
b Then, around 10 000 BCE, we began to farm. And the population began to rise.
c For most of our history we made tools from stone. We were in the Stone Age!
d But around 3300 BCE, we learned how to make bronze. The Bronze Age had begun.
e Around 1200 BCE we learned how to produce iron. It was great for farm tools and other things. The Iron Age began.
f The Industrial Revolution began in Britain, around 1760. It spread to other countries. (Some are still industrialising!) It kicked off a population explosion.
g Still rising!
World population (billions)
0
1
2
3
4
5
6
7
8
10 000 BCE
8000
6000
4000
2000
1 CE
1000
2000
(year)
7 billion in 2012
6 billion in 1999
5 billion in 1987
4 billion in 1974
3 billion in 1959
2 billion in 1927
1 billion in 1804

Your turn

1 Write out this sentence, unjumbled! *The year are than world's there rises each because more births population deaths.*

2 a Define the term *life expectancy*.
  b Look at photo **A**. Explain why the development of vaccines has helped to increase life expectancy.
  c Give one other reason why the life expectancy of a new baby in Britain today is higher than in 1700.
  d Life expectancy varies around the world. For example it is 65 in India. Suggest one reason why it varies.

3 Graph **B** shows how the world's population has grown.
  a Describe the shape of the graph.
  b The population of hunter-gatherers was low. (See note a.)
    i What are *hunter-gatherers*? (Glossary?)
    ii Suggest reasons for their low population. (What risks did they face? What weapons did they have?)

4 We began to farm around 10 000 BCE. (See note b on **B**.) Suggest one reason why this helped the population to grow.

5 Look at note e on **B**. Iron was used to make ploughs and other farm tools. Explain why this helped the population to grow.

6 a What was the *Industrial Revolution*? (Glossary?)
  b *In the Industrial Revolution, people moved from farms to towns, to get paid work in factories. Their diets improved. They met lots of people, and marriage rates increased.* Suggest two reasons why the Industrial Revolution led to a sharp rise in the world's population.

7 a When did Earth's population reach 1 billion?
  b About how many years did it take to go:
    i from 1 billion to 2 billion? ii from 3 billion to 6 billion?
  c Can the world's population just keep on rising? Decide, and give your reasons. Write at least 10 lines.

2.2 So where is everyone?

Here you'll see how we are spread unevenly around Earth – and explore the reasons why.

From empty to crowded

Some places, like Antarctica, are empty. People only visit.

Some are **sparsely populated**. For example, much of Australia.

Some are crowded, or **densely populated**. Like Mumbai in India.

Earth's population distribution

We began spreading around Earth about 60 000 years ago. Map **A** shows where we live today. The deeper the shade, the more people there are.

A

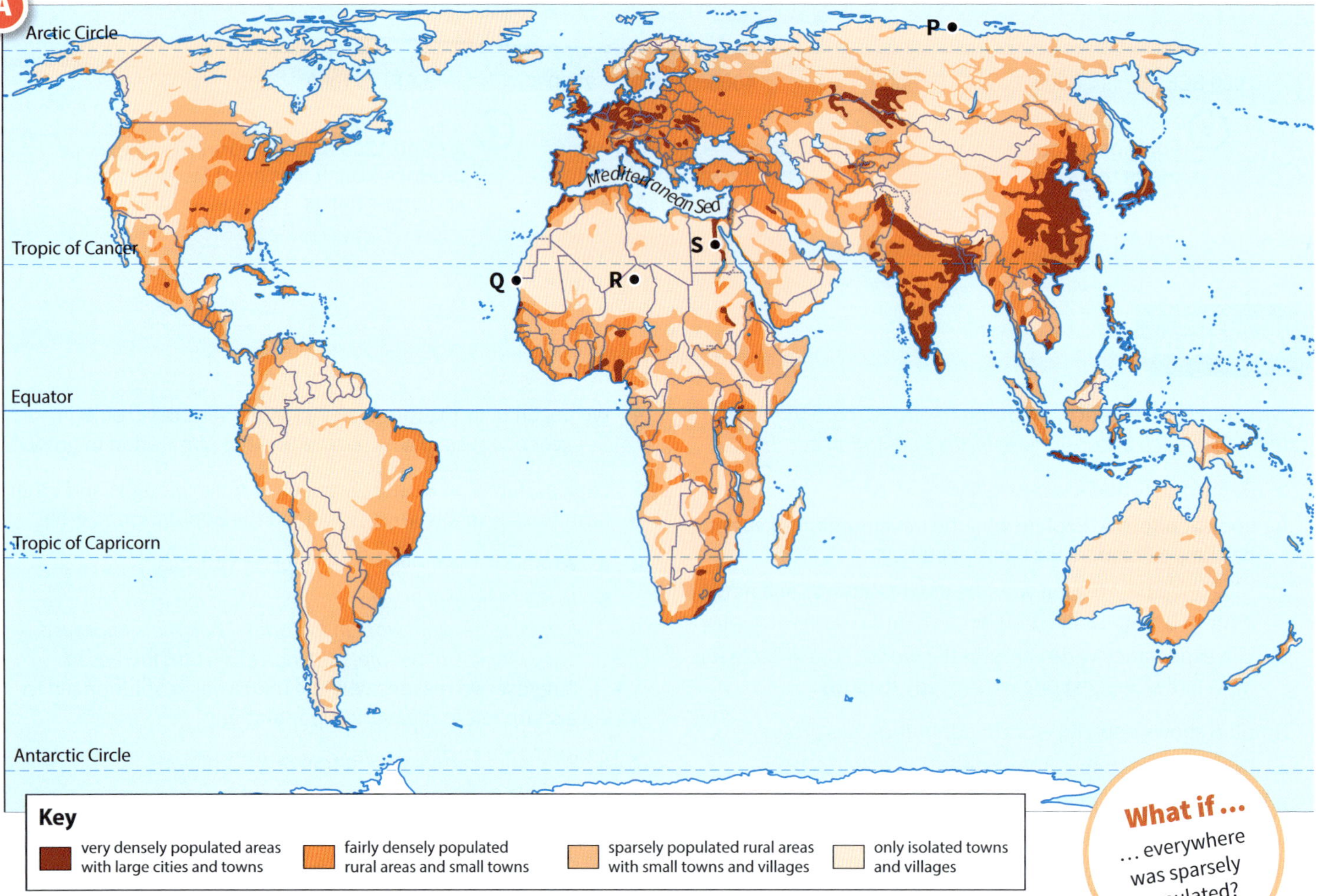

Key
- very densely populated areas with large cities and towns
- fairly densely populated rural areas and small towns
- sparsely populated rural areas with small towns and villages
- only isolated towns and villages

What if …
… everywhere was sparsely populated?

22

Population

Why is the population distribution so uneven?

As map **A** shows, some regions and countries have lots of people and some have very few. Many factors affect population distribution. Look at these:

Climate Some regions are too cold to live in – for example Antarctica. Some are mostly too hot and dry – such as the Sahara. Some climates are better than others for farming. (**C** shows what crops need.)

Fertile farmland Remember, our ancestors were farmers. Today, nearly a third of the world's population still lives by farming.

Relief Soil gets washed off steep slopes. And the higher you go, the colder it gets. So low flat land is best for farming – and easier to build on.

Rivers People need water for drinking – and for watering their crops, if there is not enough rain.

Coast A coast allows trade by ship with other countries. In fact 90% of the world trade in goods is still by ship! A coast also means fish to eat.

Other natural resources For example coal, iron ore, oil, gas. As countries began to industrialise, access to coal and iron ore were especially important. Factories set up close to these resources.

Jobs Today, most of us do not farm. We have all kinds of jobs. Some of us work in factories and in construction (building things). Over 45% of us provide **services**. Most jobs are in towns and cities, so these grow and spread.

▲ Look at this satellite image. S matches S on map **A**. It's in Egypt, which is mostly desert. But people grow crops at S because they can use the River Nile for irrigation.

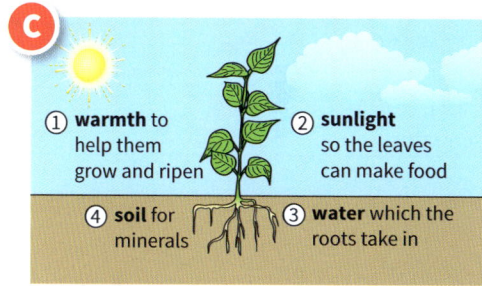

▲ What crops need, to grow.

Your turn

*You will need to compare map **A** with other maps for many of these questions. The other maps are on pages 88 – 89 and 140 – 141.*

1 Define these terms. Use the glossary only if you get stuck!
   a densely populated   b sparsely populated
   c population distribution   d population density

2 The deeper the colour on map **A**, the higher the population density. Which two *continents* have the largest areas of very high population density?

3 Name two *countries* that appear to be:
   a very crowded, overall   b very lightly populated

4 Look at the factors that affect population distribution, above.
   a Which factors are about *physical geography*? (Glossary?)
   b Give two reasons why population density is usually lower in mountainous areas than in lower flat areas.
   c Define the term *natural resource*.
   d Are rivers a natural resource? Explain your answer.

5 a Using map **A**, describe the population distribution:
      i along China's coastline
      ii around the coast of the Mediterranean Sea
   b Give two reasons why people began to settle along coasts.

6 a Find P on map **A**. It's on a coast – but nobody lives there. Suggest a reason. (Page 88?)
   b Suggest a reason why few people live on the coast at Q.

7 a R on map **A** is in northern Niger, which lies in the Sahara. Give *two* reasons why very few people live at R.
   b The Sahara extends into Egypt. Much of Egypt is empty. But there is a wiggly strip of high population density, which includes S on map **A**.
      Explain this strip of high population density. **B** and **C** will help.

8 Greenland is about 9 times larger than the UK. But it has fewer than 60 000 people. The UK has over 67 million.
   a Explain why the population of Greenland is so low.
   b Suggest three reasons, to do with physical geography, why the UK has quite a high population for its size.

9 Now, using map **A** and the maps on pages 88 – 89 and 140 – 141 (or an atlas) give your own examples of:
   a a country with a low population due to altitude
   b a country with a low population due to climate
   c eight countries with capital cities on the coast

23

2.3 Population growth around the world

Here you'll learn how the population is growing faster in some countries than others ... and why.

Population growth around the world

Earth's population is growing at about 1.1 % a year. That might not seem much. But it means Earth is now gaining over *80 million people* a year. Look at the figures on the right, for 2019.

But population is not growing at the same rate everywhere. It is even falling in some countries. Look at map **A**, for 2019. Check the key.

Earth's population in 2019

At start of year: 7631 million
At end of year: 7713 million
Increase: 82 million
So growth rate: 1.1 %

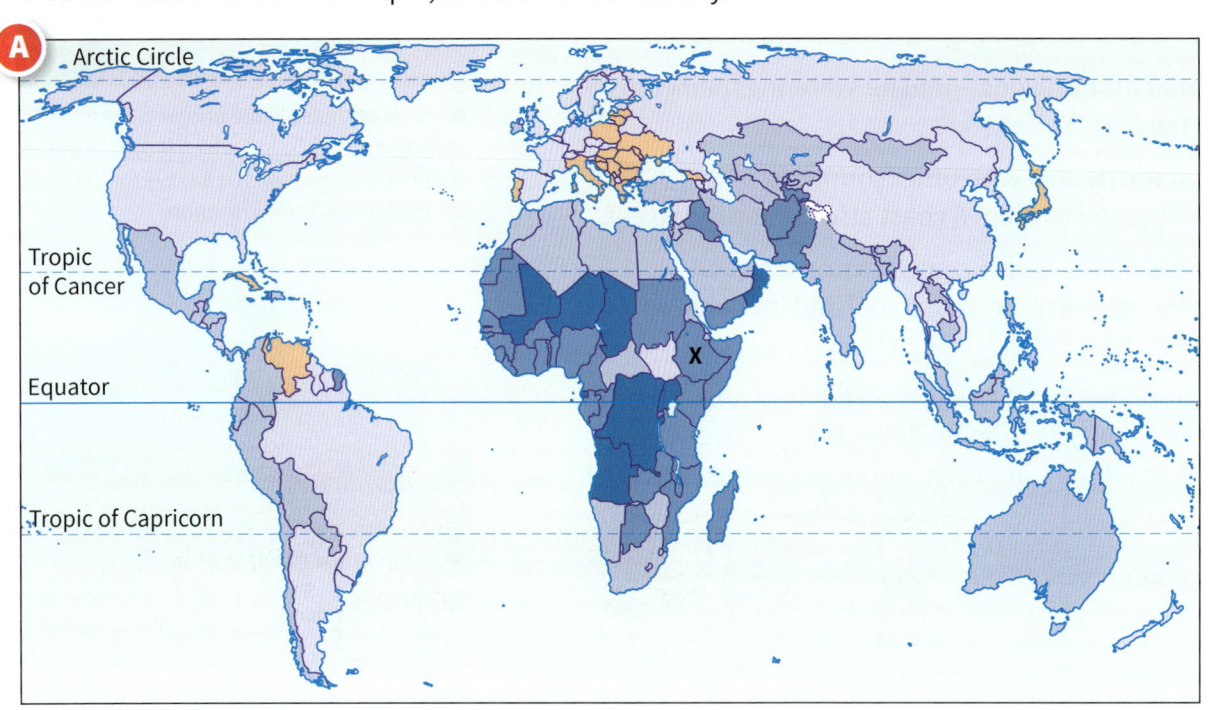

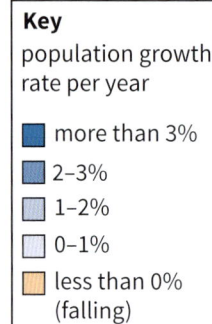

Why the differences in growth rate?

It's mainly about how well off people are!

The wealthier a country gets, the more **developed** it usually becomes. Healthcare improves. A sewage disposal system and a clean water supply are installed.

So life expectancy rises too. More babies survive to be adults and have children of their own.

Does this mean the population rises faster? No! Because, as people get better off, *women tend to have fewer children*.

In other words the **fertility rate** falls. The fertility rate is the average number of children per woman.

Look at graph **B**. Wealth is on the *x* axis, expressed as **GNI per person** in dollars. Fertility rate is on the *y* axis.

As wealth rises, note what happens to the fertility rate.

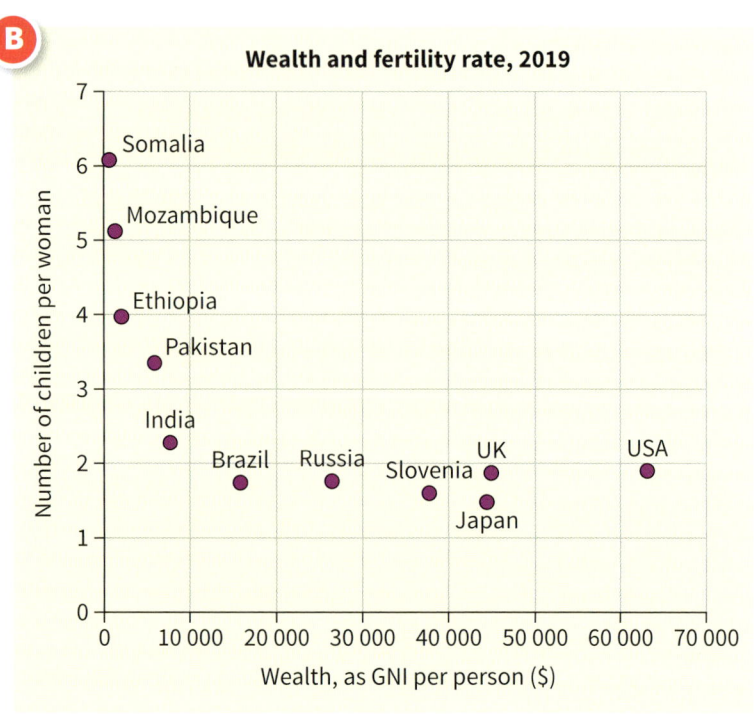

Population

Explaining fertility rates

In poorer, less developed countries

In poorer countries, many children work to support their families. Later, they'll look after their ageing parents. So a big family is welcomed.

Girls are often poorly educated. They drop out of school early. They are expected to marry young and have lots of children.

Many young women have no access to advice about planning a family – and little control over how many children they have.

In richer, more developed countries

In richer countries, women are better educated. They go out to work, and put off being mums till later. This means they have fewer children.

Parents want to give their children a good life. This can cost a lot. (And children don't earn!) So parents opt for smaller families.

Some parents keep their families small in order to help the planet! (As you'll see later, our population rise is harming other life on Earth.)

Your turn

1 This question is about map **A**.
   a Which continent had more countries with:
      i   a population growth rate of over 2%?
      ii a falling population?
      The small map on page 140 may help.
   b What was the population growth rate in the UK?
   c Name three countries where:
      i   the population grew at over 3% a year
      ii the population growth rate *fell*
2 Define these terms. (Glossary?)
   a development   b fertility rate   c GNI per person

3 Look at graph **B**, for 2019.
   a Which country was wealthier that year, Ethiopia or Japan?
   b About many children on average did each woman have:
      i in Ethiopia?   ii in Japan?
   c Explain why fertility rates are not usually whole numbers.
   d What was the fertility rate for the UK? (Include the unit!)
   e Using **B** to help you, describe what happens to fertility rates as wealth rises.
4 Give two reasons why:
   a women in poorer countries tend to have more children
   b women in richer countries tend to have fewer children

2.4 How is the UK's population changing?

 Around 2000 babies will be born in the UK today. But what's happening to the population overall? Find out here.

The UK's population is growing

You saw earlier that the world's population is growing. And so is the UK's!

Modern humans like us (*Homo sapiens*) first arrived here about 12 000 years ago, as the ice sheets from the last ice age melted.

For ages our numbers grew very slowly. By 1000 CE there were only about *1.5 million* of us. But in the 18th century the population began to shoot up. Now we are *over 67 million*.

Graph **A** shows how the population grew from the year 1000 to 2019.

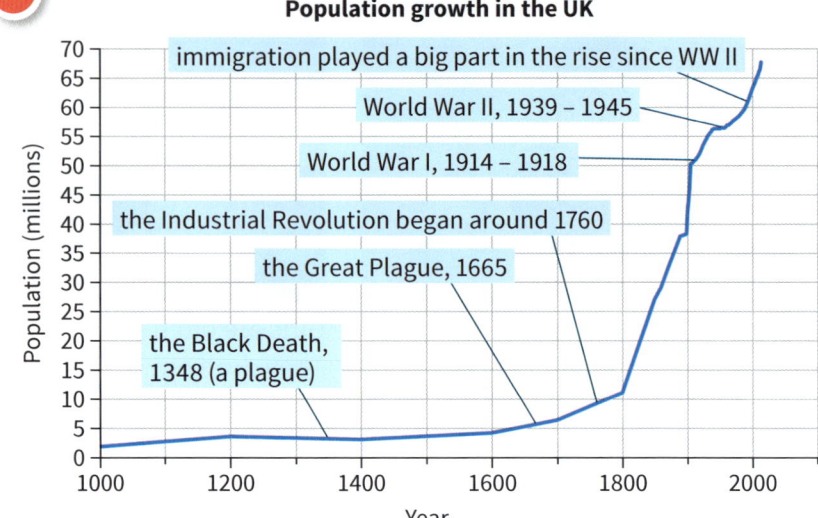

Why is the population growing?

The UK's population is growing for three reasons:
- There are more births than deaths each year. This is **natural increase**.
- People are living longer. (So they are part of the population for longer.)
- People are moving here from other countries. They are **immigrants**. People leave the UK too, to live in other countries. They are **emigrants**. But overall there are more immigrants than emigrants.

What if… … there were more deaths than births?

Did you know?
- The UK has a census (population count) every 10 years.

You calculate the *change* in the population in a year like this:

> **population change in a given year** = (number of births − number of deaths) + (number of immigrants − number of emigrants)

▲ Before the Industrial Revolution, most people lived by farming. Improvements in farming supported the population rise.

▲ During the Industrial Revolution, people left farming for factory work in towns and cities. Britain's population rose sharply.

▲ Children as young as five worked in coal mines and factories, to support their families. (Like in poorer countries today.)

Population

Birth and death rates

You saw in graph **A** that the UK's population is rising.

It would rise even faster except that … women are having fewer babies, overall. (See page 25 for reasons.)

The **birth rate** is the number of live births per 1000 people. It falls when women have fewer babies.

The **death rate** is the number of deaths per 1000 people. It falls when people are living longer. (In other words, when their life expectancy is rising.)

Graph **B** shows the UK's birth and death rates from 1960 to 2019. The lines zig-zag – but the *trend* is downwards.

How old is everyone?

The government needs to know how many people there are in the country – and how old they are. This helps it to plan ahead. For example to plan more help for the elderly.

A **population pyramid** shows a country's population divided into different age groups.

C shows the population pyramid for the UK in 2019.

Boys aged under 10 made up 6.4% of the population in 2019. Girls under 10 made up 6%. So children under 10 made up 12.4 %. (6.4 + 6 = 12.4)

A very small % of the population was aged over 100! So small that it rounds off to 0 %.

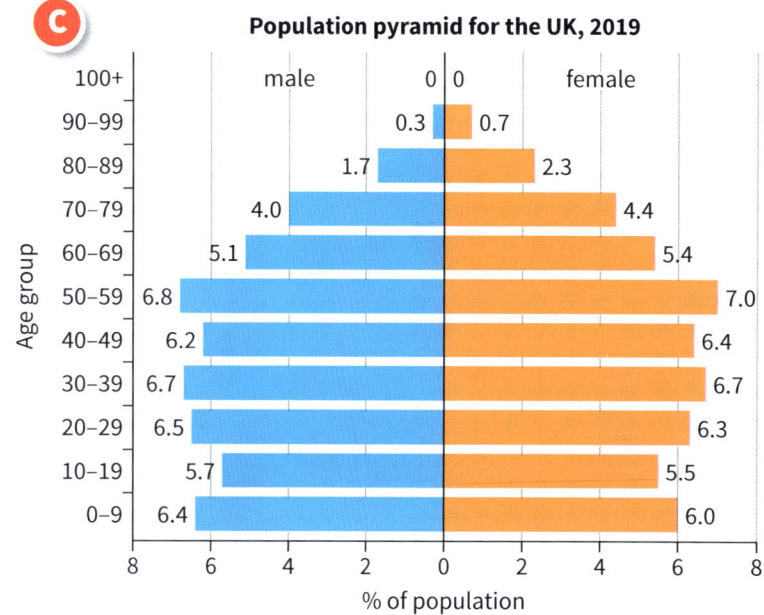

Your turn

1. Graph **A** shows the UK's population for 1000 to 2019 CE.
   a. Describe the shape of the graph.
   b. Now compare it to graph **B** on page 21, for the world's population. In what ways are the graphs similar?

2. For graph **A**, about how big was the population in:
   a. 1600?   b. 1800 ?   c. 2000?

3. The Industrial Revolution began in the UK.
   a. Describe its impact on the UK's population.
   b. Explain why this factor helped the population to grow:
      i. The factory workers received regular wages.
      ii. Steam engines were developed to drive farm machinery, so more food was produced.

4. Country X had 5000 people at the start of last year. During the year, 60 babies were born live, 20 people died, 40 people immigrated into X, and 10 people emigrated from X. Using the equation on page 26 to help you, calculate the population of X at the end of the year. (Not just the *change*!)

5. Look at graph **B**, for the period 1960 to 2019.
   a. The birth rate fell overall, across that period. Suggest two reasons why fewer babies were born.
   b. The death rate fell too. Suggest one reason.
   c. Identify one year on the graph when there was little or no *natural increase* in the population. (Glossary?)

6. Look at the population pyramid in **C**, for 2019.
   a. What % of the UK's population was aged 10 – 19?
   b. More secondary schools would be needed over the next few years. Use **C** to explain why.
   c. i. Which age group had most people?
      ii. What effect might it have on the UK, when all the workers in this age group retire?
   d. The group aged 90 – 99 was quite small. Why?
   e. *Women tend to live longer than men*. Give evidence from **C** to support the statement in italics.

2.5 What is our impact on our planet?

This unit is about the impact of our growing population on planet Earth.

Our numbers are rising fast

Just think. By this time tomorrow, Earth will have about 225 000 more humans. By this time next year, there will be over 80 million more of us.

The population will keep rising like this for some time.

Where will we all fit? Will we have enough food? Will we have enough water?

Our demand for resources

As our numbers rise, so does our demand for Earth's **resources**. We are hungry for them.

For example, the more of us there are …

▲ More new arrivals …

… the more food is needed, and water for drinking and washing …

… the more homes and roads we need, and larger settlements …

… the more fuel we need, for cooking, heating, transport, industry …

… which means more land cleared for farming, and more water pumped from rivers and aquifers for irrigation and water supply.

… which means even more land cleared. More rock used for building material. More metal ores mined to make things like cars and fridges.

… so we've been extracting and burning fossil fuels – coal, gas, oil – and petrol and diesel (from oil) at an increasing rate, for decades.

We even compete with each other for resources. There have been wars over land, and oil! One day, quite soon, there may be wars over fresh water.

Population

Our impact on Earth

We are a clever species. We make use of Earth's resources in all kinds of ways. But it's not all good news.

In our hunger for resources, we destroy the habitats of other living things. Dozens of species go extinct every day. Even pandas are at risk!

The more resources we use, the more waste we create. We dump it on land and at sea. Some will hang around for centuries.

But our biggest impact is climate change. Scientists say we're causing it, mainly by burning fossil fuels. It affects all living things, everywhere.

Those are just some of the ways we harm our planet, and other species. The larger our population, the greater our impact will be.

But there is hope …

We are clever. We can make changes. So there is hope for the future.

- We are learning that we must live in a more **sustainable** way, which does not harm us or other species, and is not wasteful.
- We are trying to protect at least some species that are under threat.
- We are trying to limit climate change, as you'll see in Chapter 6.
- Experts predict that the world population will eventually fall, as women have fewer children. Find out more in the next unit.

What if… …we ran out of land to live on?

What if… …everything was extinct except us?

Did you know?
- If everyone lived like people in the richer countries, we'd need the resources of 4 Earths to support us.

Your turn

1 a Define *resources*. (Glossary?)
   b These are some of Earth's resources:
      water   soil   wood   metal ores   oil
      Choose just two, and say how you depend on them. Give your answers as spider maps, like this:

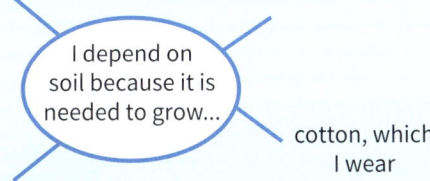

2 Look at the resources listed in question **1b**.
   a Which of them would you die without? Explain why you chose this / these, and not the others.
   b Which could we run out of? Explain your choice.

3 *Pandas feed on bamboo. Once there were plenty of pandas in the bamboo forests of Vietnam and China. But the forests were cut down for farming, and to make way for towns. Now there are fewer than 2000 pandas left on Earth.*
   Think about this. Then make up a conversation between two pandas, about the impact of humans on pandas.

4 Suppose the human population doubles in the next 50 years. What problems is this likely to cause:
   a for humans like you?
   b for other animals?
   (Don't forget things like household rubbish, and sewage.)

5 Look at this idea.
   Do you agree with it?
   Decide, and give your reasons.
   Write at least eight lines!

29

2.6 What does the future hold?

How many humans will Earth have by the end of this century? This unit looks ahead.

Predicting the world's population growth

In 1950, Earth's population was 2500 million (or 2.5 billion).
In 2000, 50 years later, it was 6100 million (or 6.1 billion).
What will it be by the year 2100? Graph **A** shows what experts predict.

A

[Graph showing Total population (billions) on y-axis from 0 to 14, and Year on x-axis from 1960 to 2100]

Why ...
... do people try to predict population growth?

▲ It is predicted that India will have 1.7 billion people by 2050, and be the most populous country in the world.

Look at the dashed line. Experts think that the population will keep rising until around 2100 – but the rate of growth will slow down. (The line will flatten out.) In 2100 it will be around 11 billion. Then it will start to fall.

It is hard to predict so far ahead. The dashed line is only an estimate. But experts are 95% sure that the growth will lie within the green area.

Why might the population fall?

Two main factors drive population: **life expectancy** and **fertility rate**.
Let's compare them:

Why ...
... can't we live to be 158?

B

Life expectancy: how many years a new baby can expect to live	Fertility rate: the average number of children per woman
As people live longer, the population grows.	As women have fewer children, the population falls.
Life expectancy is rising all over the world, as countries develop.	The fertility rate is falling around the world, as countries develop.
It rises because of better food, a clean water supply, safe sewage disposal, better healthcare.	It falls as women get better educated. They go out to work, delay having children, and choose how many to have.
It will level off – unless scientists find a way to stop our bodies from aging!	It may keep falling, since having children is a matter of choice.

As life expectancy levels off, the fertility rate becomes the key factor.
If the global rate falls below 2.1 children per woman, the world's population will fall.

In 1950 the global fertility rate was 4.7 children per woman. Today it's around 2.4! The rate is different in different countries, as you saw on page 24. But overall, it is falling quite fast.

Did you know?
- In 2018, the UK had 13 170 people aged 100 and over.

Population

A closer look at that rate of 2.1

A fertility rate of 2.1 children is called the **replacement fertility rate**. That's because 2.1 children replace the two parents (when the parents die).

Why is it 2.1 rather than 2? Because more babies are boys – and also some girls will die before they can have children.

Problems ahead!

In most countries the population is still rising – including in the UK. But in some, it is already falling. For example in Japan.

A rising population can lead to problems. But so can a falling population! Let's compare two countries, Japan and Ethiopia.

▲ Ah! The population of Japan is falling.

C Japan in 2019

Population: 127 million

Life expectancy: 86 years

Fertility rate: 1.4 children

Population *falling* by about 0.3% a year

Japan in 2050 (predicted)

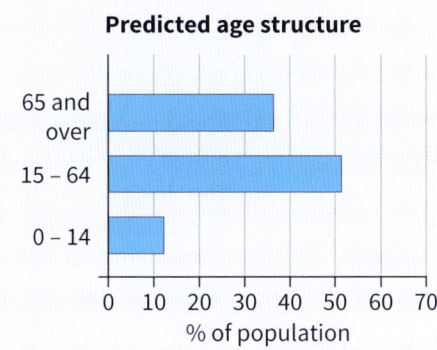

Predicted population in 2050: 107 million

There will be far more people aged over 64 than under 15. It's an **aged population**.

Just over half of the population will be of working age (15 – 64). They have to support all the rest.

When the under-15s grow up, there will be even more older people to support – and not enough workers. Japan will need immigrants.

D Ethiopia in 2019

Population: 112 million

Life expectancy: 66 years

Fertility rate: 4.2 children

Population *growing* by about 2.5% a year

Ethiopia in 2050 (predicted)

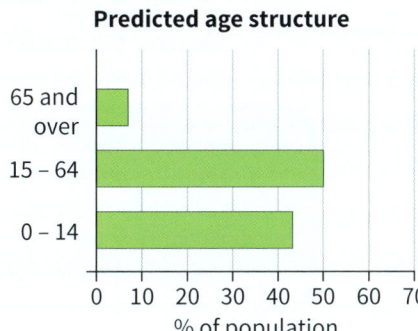

Predicted population in 2050: 189 million

Although life expectancy is rising, the % of people aged over 64 will still be low.

Over 40% of the population will be under 15. It's a **very young population**.

Because there are so many under-15s, they may find it hard to get work when they grow up. Some may have to emigrate.

Your turn

1. **a** What was Earth's population in 2000?
   b By how many billion is the population likely to increase in the 21st century? (Use the dashed line in graph **A**.)
   c What is the reason for the green area on graph **A**?

2. Look at **B**. What impact does this have on population?
   a a rising life expectancy   **b** a falling fertility rate

3. Give the *main* reason why the world's population is likely to start falling, around 2100.

4. 2.1 children per woman is the *replacement fertility rate*. Explain why: **a** it is called *replacement*   **b** it is not 2.0

5. Find evidence from **C** on page 27 that more babies are boys.

6. Compare Japan and Ethiopia in **C** and **D** above, for 2019.
   a In which country could a newborn baby expect to live longer – and how many years longer?
   b What was the average number of children per woman:
      i in Ethiopia ?   **ii** in Japan ?
   c Which country is more developed? Give evidence to support your answer.

7. Describe one challenge predicted for 2050 for:
   a Japan   **b** Ethiopia

2 Population

How much have you learned about population? Let's see.

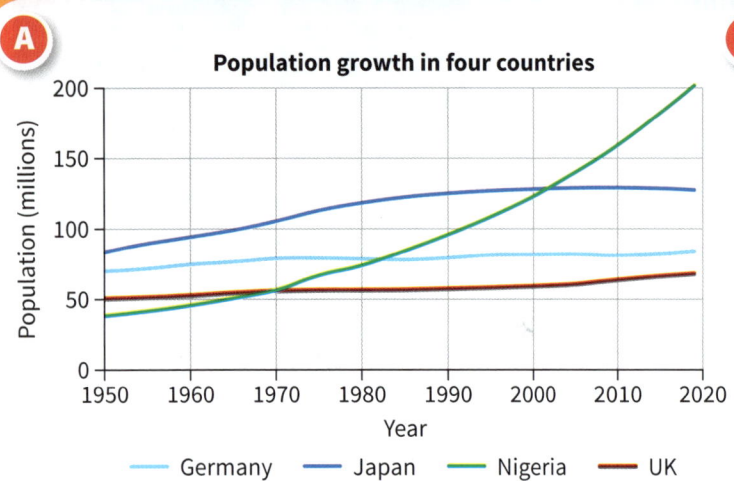

1. Graph **A** shows how the populations of four countries changed over the period 1950 – 2019.
   a. By 2019, which of the four countries had:
      i. the largest population?   ii. the smallest population?
   b. i. In which country did the population rise fastest?
      ii. State one problem that a fast-rising population may cause for a country. (Think about what people need.)
   c. In which country was the population in decline by 2019?
   d. The population of Germany was declining by 2010. (The fertility rate was 1.4.) Then it began to rise again, partly because Germany took in refugees from Syria and other countries.
      i. Define *refugee*.
      ii. Germany makes and exports far more goods than any other European country. Suggest a reason why Germany is anxious to avoid population decline.
   e. The fertility rate for Nigeria is 5.4 children per woman. Explain the link between this fertility rate and the population growth in Nigeria.

2. As you saw in **B** on page 21, Earth's population is rising fast.
   a. Which of these best describes Earth's population today?
      over 70 billion   about 77 million   near 8 billion
   b. Population changes as birth and death rates change. Define: i. birth rate   ii. death rate
   c. Graph **C** on the right shows the global birth and death rates from 1950 to 2050. Define *global*.
   d. i. Describe how the global birth rate is changing.
      ii. Give two reasons for this change.
   e. The global death rate fell between 1950 and 2020. Suggest one reason for this fall.
   f. The global death rate is predicted to rise after 2030 for several reasons, including the likely impact of climate change. If it rises above the global birth rate, what will happen to Earth's population?

3. **B** shows a very small sample of the UK's population, having fun at the seaside. **A** has a graph line for the UK.
   a. About how large was the UK's population in 2019?
      around 8 million   around 68 million   around 8 billion
   b. The UK's population has grown since 1950 through *natural increase*, and because there have been more *immigrants* than *emigrants*. Define the three terms in italics.
   c. Name two natural resources the UK will need more of, if the population continues to grow.
   d. Hedgehogs hide in hedgerows. Pesticides can kill them. They can't run fast. There were around 30 million hedgehogs in Britain in 1950. There are around 1 million today.
      Suggest two reasons for the decline in the hedgehog population, linked to the rise in the human population.

4. *Some geographers say we need to reduce Earth's population urgently, to below today's level.*
   Why do they say this? And what could be the advantages and disadvantages of a smaller population?
   Try to write at least two-thirds of a page in your answer.

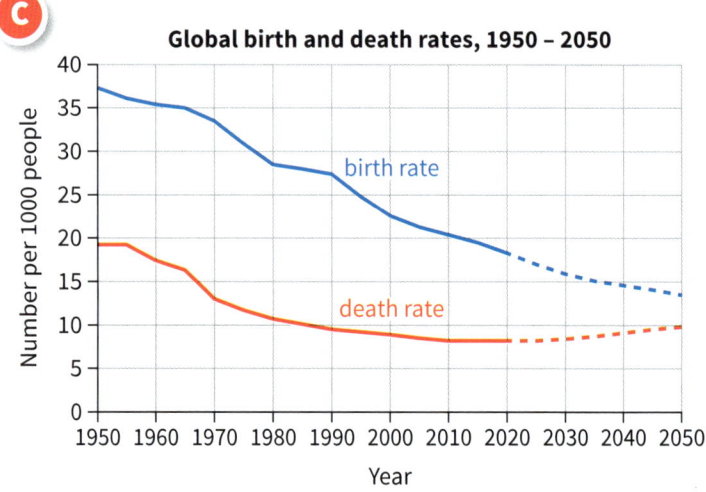

3 Urbanisation

3.1 How did our towns and cities grow?

 Today, more than half of all humans live in towns and cities. How did this come about?

> **Did you know?**
> - In the UK, more than 8 out of 10 people live in urban areas.

Once upon a time …

We take towns and cities for granted. But 10 000 years ago, there weren't any!

Once upon a time, our ancestors lived by hunting, and eating wild fruit and seeds. They were always on the move, looking for food. But then …

… about 12 000 years ago, they began to plant seeds and rear animals. (The first animals they reared were sheep.) It was the start of **farming**!

Farming meant people could settle in one place. So over time, clusters of dwellings grew. It was the start of **settlements**.

As farming developed, farmers grew extra to sell. So now, not everyone needed to farm! Markets began. Villages grew around markets.

Some villages had advantages over others (for example, easier to reach from many directions). These grew into **towns**, with many services.

The Industrial Revolution began in the 18th century. Factories sprang up in or near towns with access to coal, water, and other things they needed.

People poured in from rural areas, hoping for work in the new factories. The towns exploded in size. Some grew into cities.

Industry spread to other European countries, and the USA. So their towns and cities grew rapidly too. New forms of transport helped.

Today, towns and cities are growing fastest in Africa and Asia. People are pouring in from rural areas, as they once did in the UK – but even faster.

▲ A rural area in Japan. A rural area is mainly countryside, with farms. But it may have villages and small towns.

▲ Tokyo, Japan's capital city. It is the world's largest urban (built-up) area, formed by several urban areas joining together.

It's urbanisation!

As the Industrial Revolution spread, the populations of towns and cities increased. This is still going on.

- Across the world, every week, around 3 million people move from rural to urban (built-up) areas.
- At the same time, lots of new babies are being born in the urban areas.
- So the % of the world's population living in urban areas is increasing. This process is called **urbanisation**.

Look at **C**. It shows the rise in world population, split between urban and rural, with predictions to 2050.

In 1500, about 4% of us lived in urban areas. In 2007, half of us did. We expect the 'urban' share to keep rising.

So urban areas are spreading …

As their populations grow larger, urban areas spread. They swallow up nearby towns, and eat into rural areas.

A city with over 10 million people is called a **megacity**. In 2019 the world had 33 megacities. They were formed by urban areas spreading, and fusing with other urban areas. The biggest is Tokyo, with over 37 million people. Think about that!

C — World population, urban and rural

Did you know?

- There were some cities thousands of years ago – but only small ones …
- … because farmers could not grow enough food to support large ones.

Your turn

1. Define these terms. (Glossary?)
   a settlement   b industry   c urbanisation
2. Give reasons to explain each statement:
   a Farming was a key step in the growth of settlements.
   b The Industrial Revolution was a key factor in urbanisation.
3. Using photos **A** and **B** to help you:
   a describe two characteristics of
      i a rural area   ii an urban area
   b suggest one reason why a person might prefer to live in
      i an urban area   ii a rural area
4. **C** shows the world's urban and rural populations.
   a How is the world's population changing overall?
   b The urban share increased in the 19th century. Why?
   c Which share has been rising faster since 1900?
   d How is the rural share expected to change in the future?
   e Is the rural share likely to shrink to zero one day? Decide, and give your reasons.
5. Write a letter to one of your hunter-gatherer ancestors, describing how life has changed since he or she was around. *At least* ten lines!

3.2 Manchester's story – part 1

In this unit and the next we look at Manchester, as an example of how British cities grew.

Manchester: the growth of a British city

Many small British towns grew into cities, thanks to **industry**. Let's look at the typical steps, with Manchester as example.

1 A settlement begins – and if it's in a good location, it grows.
- Manchester began with the Romans! Around 79 CE, they built a fort near a crossing point on the River Medlock. A settlement grew around the fort.
- The Romans left, but Manchester grew as a market town, serving the rural areas around it.

2 Industrialisation arrives.
By 1760, the population of Manchester was about 17 000. Then …
- The Industrial Revolution began. New machinery was invented – including machines for spinning and weaving fibres to make cloth.
- Cotton fibre began to be imported into Liverpool. From there it could be brought to Manchester by boat. (Look at **A**.)
- So cotton mills sprang up in and around Manchester. Plus factories for dyeing and printing the cloth, and warehouses for storing it ready for export. Manchester became the world centre for cotton trading.

3 The population explodes.
- People poured in from the rural areas around Manchester to work in the mills and factories. They came from other parts of Britain too.
- Later, many arrived from Ireland, fleeing the Great Famine (1845–1849).
- In 1853, with its population over 300 000, Manchester was named a city.

4 Industry declines.
- Industry all over the UK **declined** in the 20th century – mainly because other countries began making things more cheaply.
- Manchester's mills and factories closed. This was **de-industrialisation**.
- Some people left to look for work elsewhere. Others were moved to housing estates outside the city limits. Manchester's population fell.

5 And then … regeneration!
- The revival or **regeneration** of Manchester began around 1980. It is still going on.
- The city has lots of new businesses.
- Its population has been rising again since 2001.

▲ Manchester has always been connected to Liverpool by water. First by river, and later by canal.

Why the area was good for cotton
- access to the coast
- a long tradition of home weaving
- river water, and access to coal, for the steam-driven factory looms
- soft water, good for washing and dyeing cotton cloth

Not only cotton
Other industries soon followed cotton. For example a chemical industry which started off making dyes, and an engineering industry which started off making looms.

▲ The cotton workers did 13-hour days. The clatter of the looms was deafening. The cotton dust caused lung problems. Accidents were common.

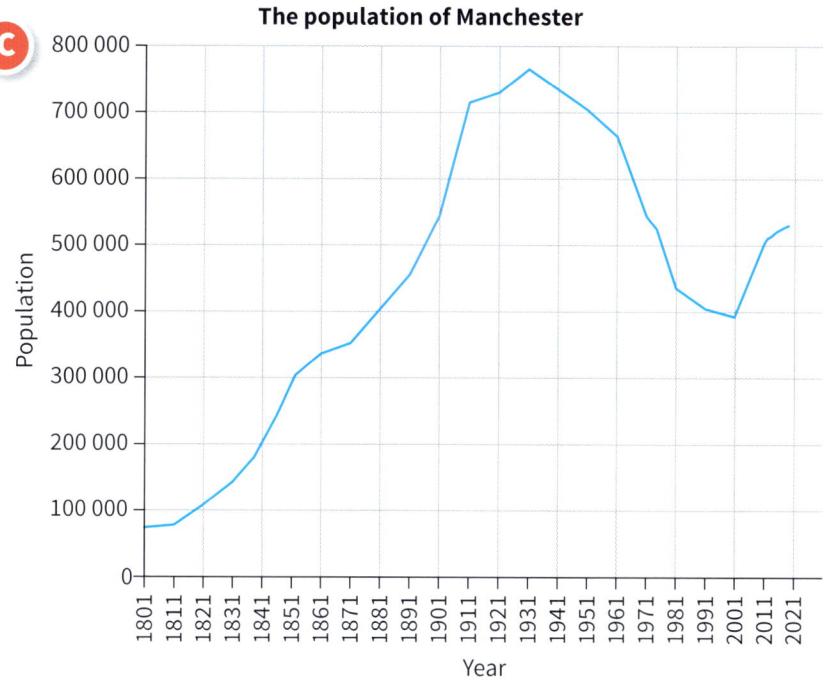

The population of Manchester

▲ The Ancoats area of Manchester in the 1870s. Slum clearance began after World War II, and people were moved to new housing estates outside the city limits.

Manchester's slums

Life was tough for the workers in the mills and factories. **Speculators** built cheap houses to rent to them, crammed into narrow streets. Most had only two rooms – one up, one down. No running water. Outdoor privies (toilets) were shared.

Over time, these areas became **slums**. Soot and fumes filled the air. Rubbish piled up everywhere. Disease was rife. Over half the children born in the slums died before age five.

Manchester spreads outwards

As the city grew, a process called **suburbanisation** took place. The better-off people moved away from the smoke and grime to areas of new homes on the edge of the city. These were the **suburbs**. Trams and buses took people into the city to work.

And so Manchester spread. It ate up countryside and villages. It fused with other towns to form a larger urban area, in an unplanned process called **urban sprawl**.

Today, this larger area, with the city of Manchester at its heart, is called **Greater Manchester**. There's a map of it on page 38.

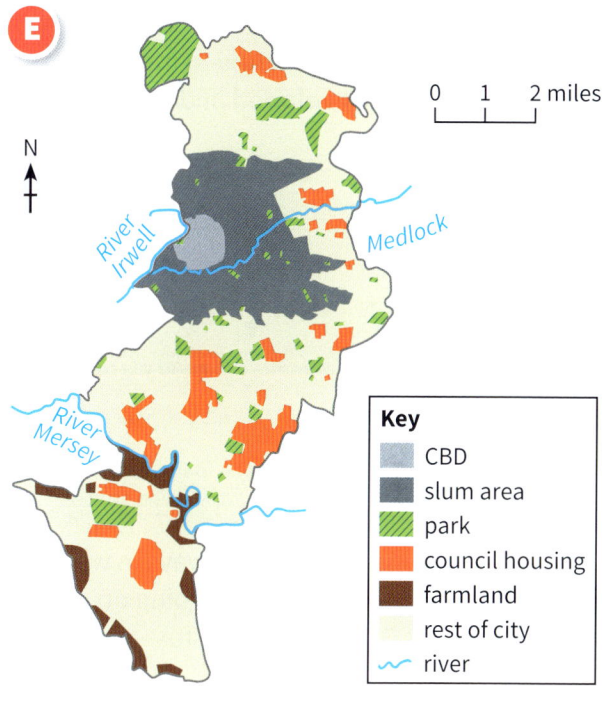

▲ Manchester in 1935. At its heart is the central business district or **CBD**. And around that, the factories and slums. By this date, the city already had council estates.

Your turn

1. Where is Manchester? Describe its location as clearly as you can. Map **A** and page 139 will help.

2. Manchester's cotton industry was hugely successful.
   a. Give: i one geographical factor ii one social factor which helped to make it a success.
   b. Its success attracted other industries. Give one example.

3. Using **C** to help you, describe and explain the change in Manchester's population between 1801 and 1931. Include some numbers from the graph in your answer.

4. As it grew, Manchester underwent these processes:
   a. suburbanisation b. urban sprawl
   Describe each process.

5. Give *two* reasons to explain why the population of Manchester declined after 1931.

6. It is 1875. You are a journalist on a visit to Manchester. Write an article about the impact of industrialisation on the city. Make it interesting and informative for your readers! **B** and **D** will help. Write *at least* 12 lines.

3.3 Manchester's story – part 2

Here we focus on the regeneration of Manchester – and why its population is growing today.

The Manchester miracle?

Look how Manchester has changed!

In 1980 Manchester was a depressed city. Mills and factories stood empty. Thousands of people had lost their jobs. The slums had gone, but much of the city was run down. People were leaving.

Today Manchester is a vibrant city, with lots of new homes and new jobs. The population has been rising since 2001. People say it's a great place to live.

So … what happened? Manchester has been **regenerated**.

How does regeneration work?

Regeneration has two key aims. Look in the two boxes below. These aims are connected. And to meet them, many different aspects of a place must be **developed** and **improved**. Look at the green labels.

Page 39 shows just some of the ways in which Manchester has been regenerated. The process is still going on. Not only within the city but throughout Greater Manchester – and in other British cities too.

Work through page 39, and then try *Your turn*.

A

▲ The city of Manchester is at the heart of Greater Manchester, a **conurbation**. The whole area is being regenerated.

B

▲ An abandoned cloth warehouse awaits a new owner. By 1815, Manchester had 1819 warehouses in the city centre!

Your turn

1. **A** shows Manchester city at the heart of a conurbation. Define the term *conurbation*. (Glosssary?)
2. Look at the building in **B**. Why is it in this state?
3. a Define the term *regeneration*.
   b State the two aims of regeneration.
   c See if you can explain how education helps in regeneration.
4. Explain clearly how this is helping to regenerate Manchester.
   a MediaCityUK
   b the free bus services in and around the city centre
   c the modern apartments being built around the city
5. Explain how these benefit Manchester's *economy*. (Glossary?)
   a its football clubs b the Parklife music festival
   (Think about what people spend money on, on days out.)
6. The population of Manchester is now rising. Explain why.
7. Compare Manchester today with Manchester in 1870. Think about jobs, air pollution, housing and so on. You can answer in any way you wish: text, a table, drawings, spider maps, a speech by the Mayor …
8. Of all the improvements in Manchester, which *two* do *you* think are the most important? Explain your choice.

Urbanisation

The regeneration of Manchester

Jobs

Many of the new jobs are in creative and digital media, and science.

MediaCityUK is in Salford, about 5 km from Manchester City Centre.

Graphene, an amazing 2-D material (1 atom thick and 200 times stronger than steel) was first isolated in Manchester University in 2004.

▲ MediaCityUK has BBC and ITV studios, and over 250 smaller media and digital companies.

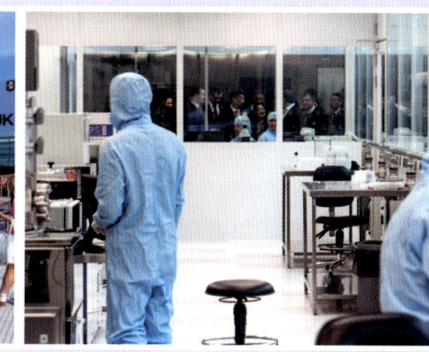
▲ Manchester aims to become a world centre for research into 2-D materials.

Transport links

People want to get around easily.

Manchester has a network of trams, light rail and buses, linking it to the rest of Greater Manchester.

It has good links to the rest of the UK by road and rail, and to the world via Manchester Airport, only 13.5 km from the city centre.

▲ One of Manchester's yellow trams. The system has 93 stops and covers 100 km.

▲ Bus services in and around Manchester city centre are free!

Housing

Everyone wants good housing.

Run-down housing is being restored or replaced. Thousands of new flats and houses are being built for the growing population.

The new ones are on **brownfield sites** – sites once used for something else – because the city wants to protect its green areas.

▲ Once a cloth warehouse, now a block of flats. Warehouses were often grand buildings.

▲ A modern apartment building in the area where the photo on page 37 was taken.

Activities, culture, shopping

There's football ... and more.

Sports venues were developed for the Commonwealth Games, which the city hosted in 2002.

Manchester Arena is one of the world's biggest indoor arenas. The city has several other venues too.

There's a great choice of shops, and cultural activities. The Lowry Museum is near MediaCity.

▲ Manchester City playing Manchester United. Their grounds are about 6 km apart.

▲ Queuing for Parklife, Manchester's biggest music festival, at Heaton Park.

3.4 Urbanisation around the world

Most people in the UK live in towns and cities. Is it the same in other countries? Find out here.

A map of urbanisation around the world

Over 83% of the people in the UK live in urban areas.

What about the rest of the world? Look at map **A** and its colour key. The deeper the shade of a country, the higher the % in urban areas.

Did you know?
- North America is the most urbanised continent, at 82%.
- Next is South America (80%).

A

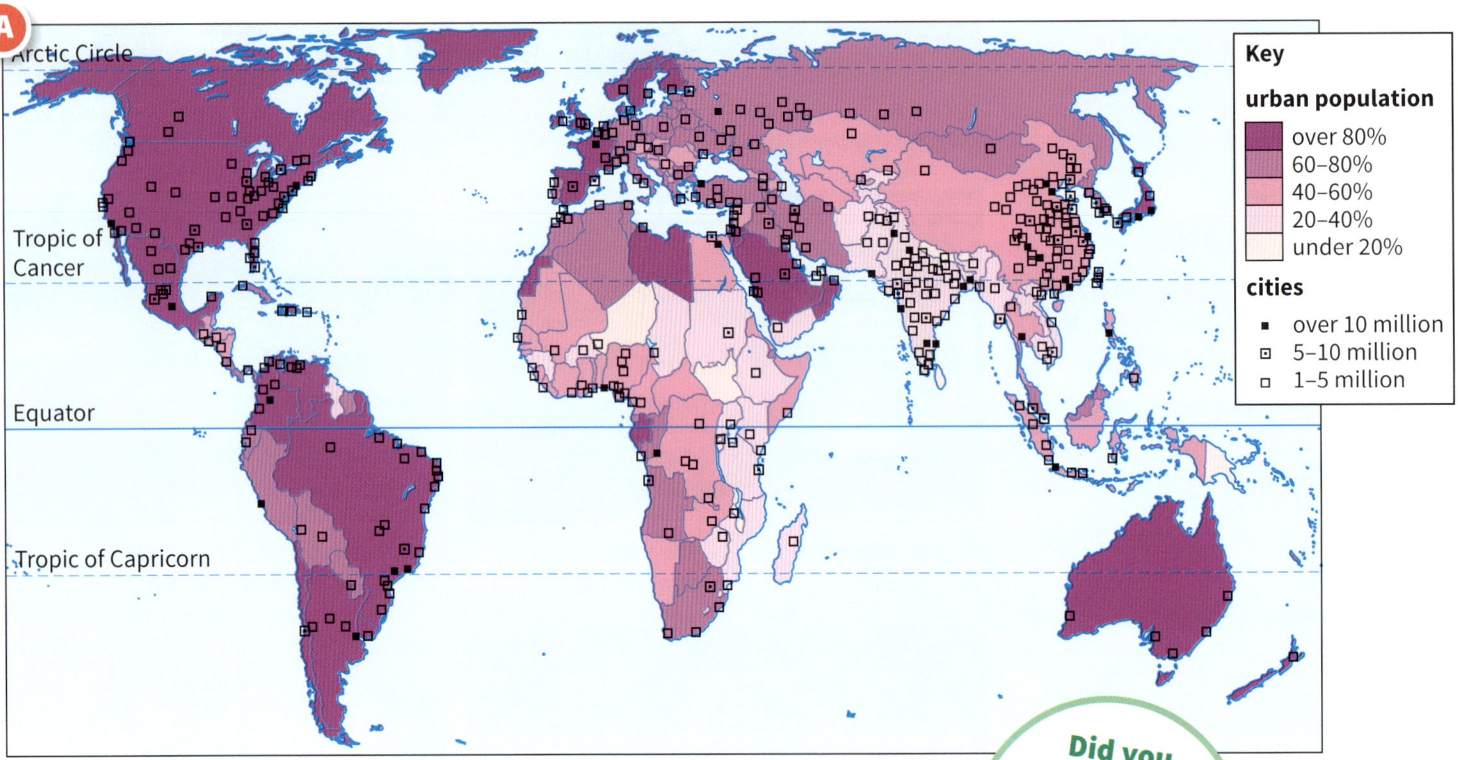

[Map showing world urbanisation with key:
urban population: over 80%, 60–80%, 40–60%, 20–40%, under 20%
cities: over 10 million, 5–10 million, 1–5 million]

Map **A** also shows the cities with at least 1 million people.

Look for the megacities, with over 10 million people. Some have far more than 10 million. For example Shanghai, in China, has around 24 million.

Did you know?
- Singapore – a small island country in Asia – is 100% urbanised.

Most cities: under 500 000 people

Map **A** shows only the cities with at least 1 million people.

In fact more than half of the world's urban dwellers live in towns and cities *with under 500 000 people*. These are not shown on the map because it would be too packed.

The map on page 139 shows the main cities in the UK. All have fewer than 500 000 people, except for London (nearly 9 million) and Birmingham (around 1.2 million).

▶ As more and more people move in, cities grow outwards – and upwards.

Urbanisation

Why is it happening?

You already learned quite a bit about urbanisation, with Manchester as example.

C summarises the process.

As this process continues across a country, the % of the population living in urban areas increases.

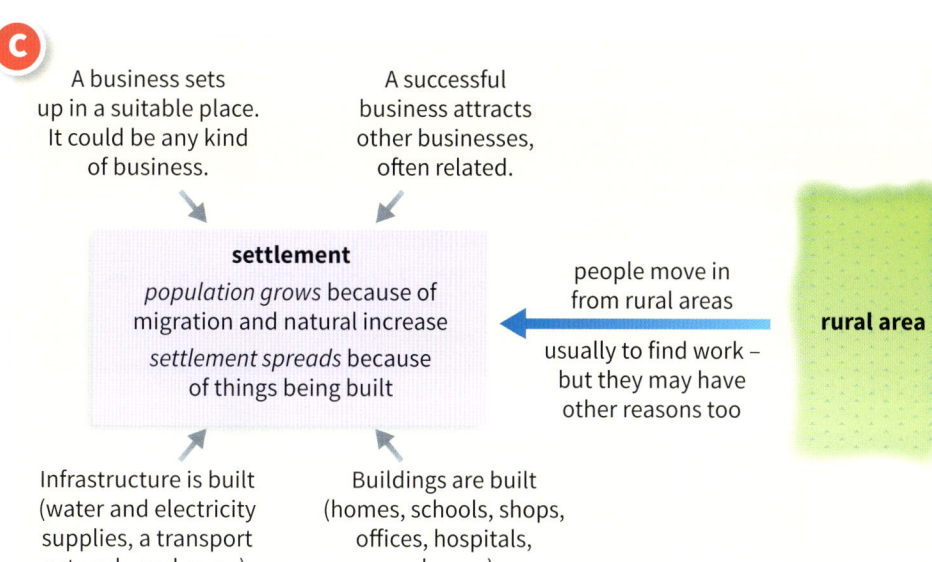

Is it linked to wealth?

Most businesses start in urban areas. When they flourish, the country grows wealthier. And more people arrive to work in them. Urbanisation increases.

Look at graph **D**.

It shows the urban % for twelve countries in 2018, plotted against their wealth. Wealth is shown as GNI per person, in dollars.

Can you spot a trend?

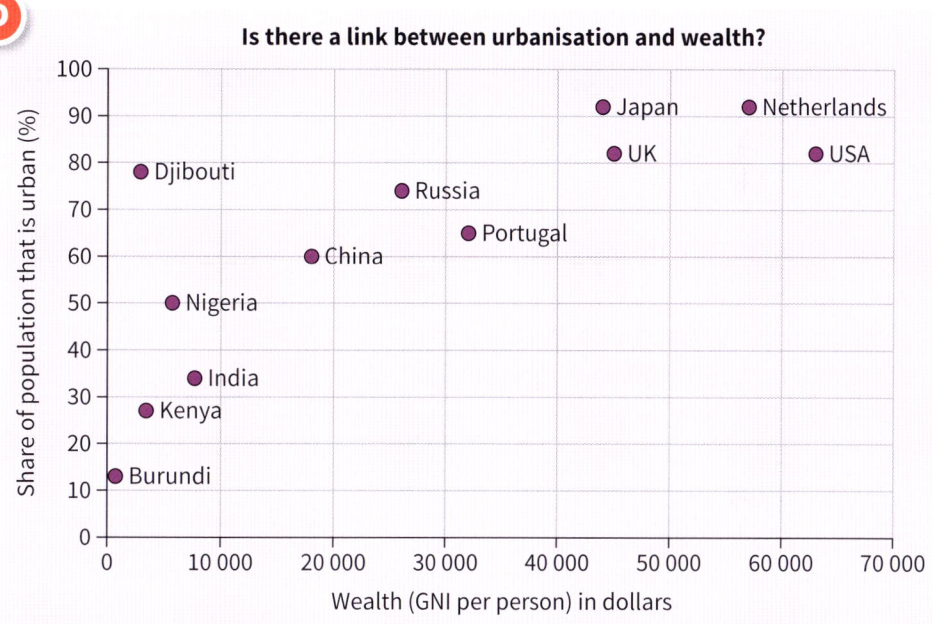

Your turn

1 Map **A** and pages 140 – 141 will help you with this question.
   a Which continent has more countries with an urban population of under 40%?
   b Name one country in South America where at least 80% of the population lives in urban areas.
   c Name one African country where:
      i   under 20% of the population lives in urban areas
      ii under 20% of the population lives in rural areas
   d Write a sentence about the level of urbanisation in:
      i   Russia   ii Canada
   e Which is more highly urbanised:
      i   China or India?   ii Australia or Italy?

2 Look at **B**. Give one advantage of building high-rise apartment blocks in a country that's urbanising.

3 Identify the main factor behind urbanisation. (**C** will help.)

4 Look at graph **D**.
   a i   Which country in the graph is the wealthiest?
      ii About what % of its population is urban?
   b i   Which country in the graph is the poorest?
      ii About what % of its population is urban?
   c Overall, as wealth increases, how does urbanisation appear to change?
   d Comment on the position of Djibouti in **D**.
   e Which country in **D** has the highest % of people in farming? Give reasons for your answer.

5 Jobs are not the only reason people move to urban areas. Factors such as climate can also play a part.
   Suggest a reason for the level of urbanisation in Djibouti. (It borders Ethiopia in Africa.) Pages 88 and 141 will help.

6 Will the data for **D** be the same for the year 2030? Explain.

3.5 Push and pull factors

 Here we dig deeper into the reasons why people move from rural to urban areas today.

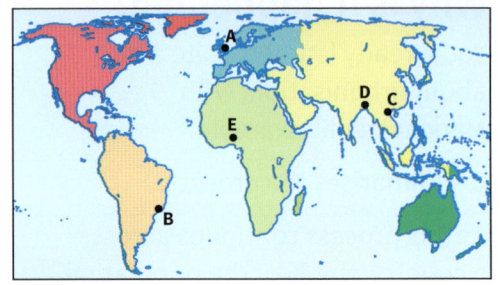

Why do people move?

Every week, across the world, *around 3 million people* leave rural areas, to move to towns and cities. Why? Look at these examples.

Sylvie, London, UK — A

I was born in a little village in Yorkshire. It's lovely – but so quiet! It was a long bus ride to school.

Then I went to art college in London. And that's where I live now. I'm a fashion designer.

I love going back home to visit. It's so peaceful. But I couldn't live there. I'd be bored! Anyway, London is the place to be, for a career in fashion. I have lots of contacts here.

Three friends, Rio de Janeiro, Brazil — B

We've been friends since we were little. We grew up in the same village. We used to talk about going to Rio, and earning loads and watching football.

We've been here for six years now. Not earning loads, but life is good. Joel (on the left) is married. Viktor (middle) saved like crazy to go to college. I work for a tour company. It's okay, Rio gets lots of tourists.

Life in the city has hassles – but it's more fun than the countryside. Give me the city any day.

Lan, Hanoi, Vietnam — C

I'm from the country. I do miss my village – but my family were so poor. I had no option but to leave the farm, to earn money.

I like the city. I like making snacks, and business is good on this street. I send money home. My parents are quite old now and they depend on me.

I am getting married to Huy soon. He's from my village too, and he's a taxi driver. We'll never be able to buy a flat – but we'll be okay.

Urbanisation

Shimaz, Dhaka, Bangladesh

I'm 19. I came to the city two years ago to help my family, because they are very poor. I work in this factory 6 days a week, making clothes for shops in the UK. Maybe for you.

I make 8400 taka a month (about £80) and send half home to mum. I'm really glad I can do this.

Life is not easy here – but I do like having freedom, and my own money. Back in my village I'd be married by now.

I have good friends in the factory. We share a dorm. On our day off we go round the market.

Osakwe and his family, Lisa, Nigeria

We're moving to Lagos next week – the whole family.

I've tried to make a go of the farm. But we had floods the last few years, and everything rotted. Now I'm broke. They say it's climate change, and it will only get worse.

I want my children to have a better life. I don't want them to be farmers. The schools here are no good. I think Lagos will be better.

My brother lives there. He says we can stay with him till we get on our feet. I don't want to be a burden to him. I'm sure I'll find a job of some sort.

Push and pull factors

The people in this unit had a range of reasons for moving to cities.

People may feel forced to move from rural areas, for example by poverty. Factors that drive you away from a place are called **push factors**.

But some places attract people because they offer a lot of advantages. For example, well-paid work, or the chance of a better education. Factors like these are called **pull factors**.

Your turn

1. What is: **a** a push factor? **b** a pull factor?

2. **a** Make a big table with headings like this:

Name(s)	Push factors	Pull factors

   b Now fill it in for the people in photos **A – E**. See how many push and pull factors you can identify.

   c In your completed table for **b**:
   - **i** how often does *poverty* appear as a push factor?
   - **ii** do any pull factors appear more than once? If yes, identify them.

3. As climate change continues, it may speed up urbanisation in a number of countries. Explain why. (Page 97 may help.) You could use terms like these in your answer:
   floods   drought   wildfires

4. Some people move the other way, from urban areas to rural areas. This change is called **counter-urbanisation**. See if you can suggest:

   a two push factors   **b** two pull factors

   that might cause people to move from a city to a rural village. The photos on pages 33 and 35 may give you ideas.

3.6 It's not all sunshine!

 Life in urban areas brings many benefits. But there are problems too. Find out more here.

> **Did you know?**
> - Burundi, in Africa, is the least urbanised country in the world …
> - … with only 13% of the population urban.

Do we like living in towns and cities?

The answer seems to be yes! Over half of the world's people now live in urban areas, most of them willingly.

Here are some of the things people like about urban living:

A Benefits of urban living

- concerts, clubs, cinemas, fun
- lots of interesting things to do
- lots of people who share your interests
- colleges, universities, all kinds of classes
- a choice of districts to live in
- great shops
- all kinds of eating places
- all kinds of work
- well-paid jobs
- hospitals, doctors, dentists
- buses, trams, trains, airports

But there are disadvantages too. Look:

B Disadvantages of urban living

- crime
- anti-social behaviour
- competition for jobs
- competition for housing
- crowds
- concrete everywhere
- pollution
- people don't know their next-door neighbours
- many people feel isolated
- more expensive than living in rural areas
- noise
- traffic congestion

44

Managing urbanisation

Imagine you are in charge of an urban area. The population is rising fast. New migrants arrive daily. Your job is to **manage** the area, to make sure it can cope.

It's a challenge. Because people need some things just to survive – and many other things to make their lives easier, and more satisying. Look at **C**.

Your job is even tougher if:

- urbanisation is happening very rapidly
- you don't have enough money to do what's needed.

Many lower-income countries face these problems.

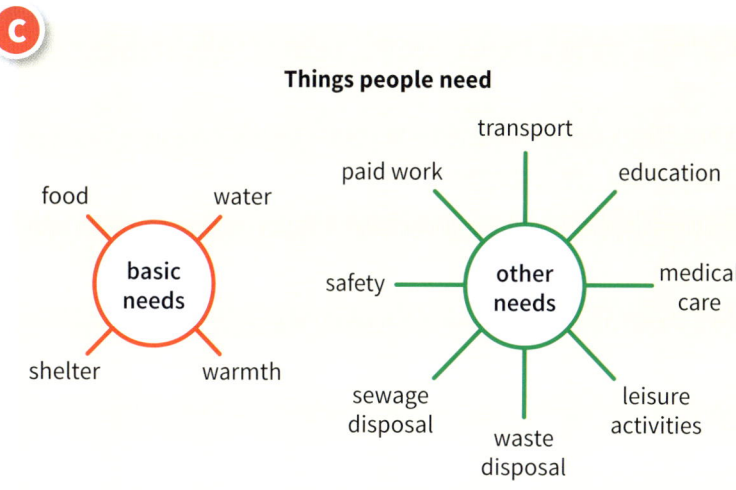

C Things people need

You may end up with slums

If you can't manage urbanisation, you may end up with **slums**. A slum is an urban area of very poor housing, with few services.

Some slums were built as cheap housing which then went downhill – like the slums in Manchester (page 37).

Other slums are **squatter settlements**. People build shacks from bits and pieces, on vacant land, without permission.

The shacks don't have an indoor water supply, or toilet. For electricity, people may hook up to illegal cables.

Around 1 in 3 of the world's urban dwellers live in slums! **D** shows data for a recent year.

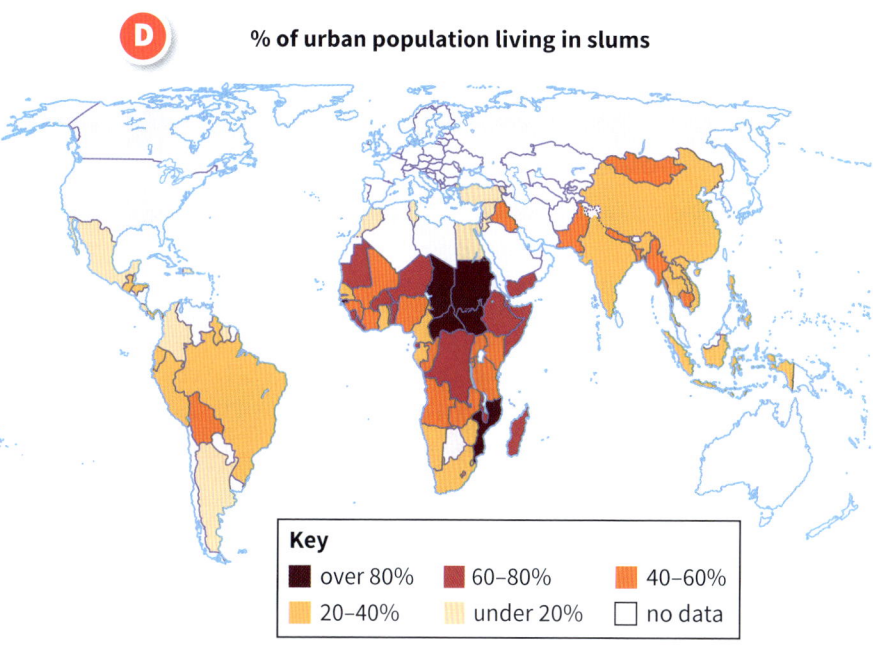

D % of urban population living in slums

Your turn

1. Look at **A**. It shows benefits of urban living.
   a. Which three would be the strongest pull factors, to attract *you* to a city?
   b. Would your answer be the same if you were 65? Explain.
2. From the disadvantages in **B**, pick out:
   a. the three you think have most impact on people
   b. the two you think most likely to push this group away:
      i. elderly people   ii. parents with young children
3. Which disadvantages in **B** may also apply to rural areas?
4. Many cities around the world have *slums*. Some are *squatter settlements*. Define the terms in italics.
5. Look at **D**.
   a. Which continent has the biggest problem with slums? Explain how you decided.
   b. Name two African countries where over 80% of the urban population lives in slums.
   c. What % of the urban population lives in slums in:
      i. Argentina?   ii. India?   iii. China?   iv. Kenya?
   d. There is no data shown on **D** for many countries, including the UK and the USA. Suggest the most likely reason.
6. Explain why slums are common in low-income countries. Use at least some of these terms in your answer:
   migrants   homes   money   manage   rapid urbanisation

3.7 Life in the slums

  Life can be difficult in the city. And extra difficult if you live in a slum. Find out more about slums here.

Osakwe in Lagos

On page 43 you met Osakwe, who was heading to his brother's place in Lagos. Lagos is the capital of Nigeria, and Africa's largest city. What did Osakwe find there?

> **My first week in Lagos**
>
> Here we are in Lagos. In Agege, with my brother Amadi. It's a slum! Never have I seen so many shacks.
>
> There are nine of us in his two rooms, so it's a bit crowded. But we have electricity! We didn't have it in my village. It's on for a few hours, most days. You never know when it will go off.
>
> There's a water tap a few minutes away. The nearest latrine is down the track. It's a hole in the ground behind a bamboo screen. It stinks, and there's always a queue.
>
> And the rubbish! People just throw rotting stuff into the gully in the lane. There are flies and dogs everywhere. I'm worried the children will fall ill. I can't afford a doctor.
>
> Amadi is a porter in the market. Eki – his wife – has a food stall. They work so hard. He thinks I should try the market too. I must find something soon, and a place of our own to live in.
>
> I must forget about school for the boy for now. I am sad about that. I hope life will be better for us here … but I'm not sure it will.

▲ What Osakwe saw when he arrived.

More about Lagos

Lagos is a megacity. Nobody knows for sure how many people live there, but it's thought to be over 21 million.

Tens of thousands of migrants arrive in Lagos every week, hoping for a better life. So the city is growing at an alarming rate.

Many have little money or education, and end up renting rooms in slums.

It's thought that about *two-thirds* of the Lagos population live in slums.

▶ People work hard in slums, trying to make a living in difficult conditions. It's tough – but they help each other.

Tackling the slum problem around the world

Lagos is not alone. There are slums in many cities in lower-income countries. They take different approaches to tackling the problem. For example:

- **new housing** In some places they build blocks of flats (like in **C**), and move people into them at low rents, and clear the old shacks away.
- **self-help** In some places, they give people building materials, and some training. Then people build their own, improved, homes.
- **better services** In some places they leave the housing in place, but provide reliable water and electricity supplies, and a sewage system, and rubbish collection. Those make life so much much better.

But it all costs money. And every day, more people arrive in the slums. The city councils don't have much money. So it's a non-stop struggle. It will take many years to solve the world's slum problem.

A quick fix?

Several slums in Lagos have been bulldozed, like the one in **E** below. The city council wanted the land for luxury apartments and shops. People were forced out – with nowhere to go except another slum. So this is not a solution!

▲ New low-rent flats for slum dwellers of Kibera in Nairobi, Kenya. The slum is in front.

▲ Slum clearance in Lagos. 30 000 people were left homeless when this slum was bulldozed. There were many protests. One day, Lagos will solve its slum problem.

▲ New homes for slum dwellers in South Africa. The roofs have solar panels, to give electricity. They also slope, to let rain water be collected. The walls are designed to keep the inside cool.

Your turn

1   **A** shows part of a squatter settlement. Draw a field sketch from **A** – without the people. Add notes. (For example about the stream, and what the shacks are made of.)

2   Imagine you are Osakwe, and you have your children to look after. What *two* things would bother you most about your first week in Agege?

3   Three ways to tackle a slum problem are listed above. Which one do you think might:
   a   i   cost least?   ii   cost most? (think!)
   b   have the greatest impact on the slum dwellers?
   c   make people feel more in control of their lives?

4   a   There are no roads in **A**, so offering self-help or better services might not be sensible options here. Explain why.
   b   What would you recommend, to help the people in **A**?

5   Look at the lady cooking in **B**. That's how she earns her living. Imagine she has a choice. She can choose:
   – self-help plus better services laid on, or
   – to move to a new apartment like those in **C**, or
   – to move to a home like those in **D**.
   Which do you think she might prefer? Explain your choice.

6   The man in **E** is a community leader. He feels angry about what happened here. Explain why.

3.8 How can we make cities more sustainable?

In this unit you will learn what a sustainable city is, and about ways to make it happen.

How cities develop

Look at **A**. It shows how cities have been developing over time, to meet more and more of our needs.

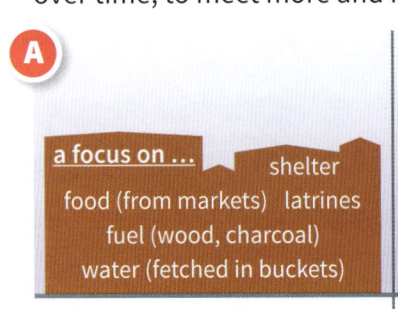

1 basic — a focus on … shelter, food (from markets), latrines, fuel (wood, charcoal), water (fetched in buckets)
providing only what people need for survival

2 improving — a focus on … schools, hospitals, shops, roads and railways, transport (buses, trains, taxis), water, electricity and gas supplies, sewage and rubbish removal
city getting organised, more of people's needs are met

3 advanced — a focus on … universities and research, finance and banking, air, sea and rail connections, internet and mobile connections, online shopping, systems run by computers
more efficient, economy growing, people getting better off

4 sustainable — a focus on … the environment, culture, greening, tackling climate change, reducing waste, leisure
reducing harmful impacts and improving the quality of life

→ development within cities

Every city – or part of a city – is somewhere along this path.

In lower income countries, much of a city may still be at stage 1 – for example its squatter settlements. Over time these too will develop. In the higher income countries, including the UK, most cities are far into stage 3. Many have entered stage 4: becoming more sustainable.

What's a sustainable city?

Cities gobble up food, clean water, and fuel. They produce billions of tonnes of waste water, sewage, rubbish, and air pollution!

A sustainable city has minimum negative impact on the environment, and minimum waste, while still offering its people a high quality of life.

We are making our cities more sustainable little by little. Look at the examples on page 49. Then do *Your turn*.

▲ Do parks and lakes improve life in the city? This is London's Regent's Park.

Your turn

1. **A** gives you a way to think about cities. Define these terms:
   a development   b sustainable city
2. List the four stages in **A**.
3. A city – or part of a city – could be anywhere along **A**, including between stages. As far as you can tell, where on **A** would this fit best? Explain each choice.
   a the area of Lagos shown in photo **B** on page 46
   b the city shown on page 33
   c the regenerated Manchester, on page 39
   d another urban area that you are familiar with
4. Page 49 shows some ways to make a city more sustainable. Identify two that:
   a improve air quality
   b reduce the need for fuel
   c reduce the need for landfill sites (glossary?)
   d help to keep people healthy
   e make life more enjoyable
   f don't cost that much
   g help in the fight to limit climate change
5. Suggest two changes that could be made in *your* local area (even if it's not in a city) to make it more sustainable.

Urbanisation

Making cities more sustainable

Transport

Cities are cutting pollution from traffic. This London bus runs on hydrogen, which gives only water when it burns. No carbon dioxide!

They are reducing car use too: for example by charging a tax to drive into the city, and setting up bike lanes and rent-a-bike stands.

New homes

More homes are being made in sections in factories, for fast assembly on site. They are well insulated, so need little heating.

Apartment blocks are being built on brownfield sites, to curb urban sprawl. Built-in gyms, cafes and supermarkets add convenience.

Waste water

Some homes have systems for reusing the waste water from baths and showers to flush toilets. This will become more common.

Household waste

Companies are under pressure to use less packaging – and make it all recyclable. Shops are selling more food with no packaging!

More rubbish is being burned in incinerators, instead of buried in landfill sites. The heat is used for electricity or to heat local homes.

Greening the city

Trees are being planted across cities. As are roof gardens, and living walls. They soak up carbon dioxide – and attract birds and insects.

More people are growing food in cities too – on balconies, flat roofs, and in community urban gardens. It's satisfying to grow your own.

3 Urbanisation

How much have you learned about urbanisation? Let's see.

check ✓

1 **A** is a recent aerial photo of part of Manchester.
   a The industrialisation of Manchester carried on for much of the 19th century.
      i Define *industrialisation*.
      ii Name Manchester's main industry in the 19th century.
      iii Describe and explain the impact of industrialisation on the size of Manchester's population.
      iv Slums developed in Manchester. Explain why.
   b The opposite process to industrialisation took place in Manchester in the 20th century. Name this process.
   c Manchester began a process of improvement in the late 20th century. It is still going on.
      i The correct name for this process is r_____?
      ii State one result of this process.

2 The river in **A** is the Irwell. In the 19th century it stank with waste from the slums and mills and dye houses. Today people canoe on it, and fish in it, and stroll along its banks.
   List four other ways in which life for people in Manchester is different now than in 1870. Pages 37 and 39 will help.

3 **B** shows how the population of Nigeria is changing. The graph is split between rural and urban.
   a Name: i a rural area ii an urban area near you.
   b i How is Nigeria's total population changing?
     ii About how big was it in the year 2000? Estimate!
   c i Define the term *urbanisation*.
     ii Give evidence from the graph that Nigeria is undergoing urbanisation.
   d Which has been growing at a *faster* rate since 1950?
      i the urban population   ii the rural population
      Explain how you decided.
   e What fraction of Nigeria's population is expected to be urban by 2050?
      i about one-third   ii about two-thirds
   f Lagos is Nigeria's capital. Its population is growing fast. State two ways in which the population of an urban area grows.
   g Give one example of: i a push factor ii a pull factor which helps to explain why people move to Lagos.

4 Photos **A** and **B** on page 46 show people in slums in Lagos.
   a Give another name for areas of illegal shacks.
   b Slums are common in cities in lower-income countries today. Give two reasons.
   c Outline two different strategies for improving the lives of people in slums.
   d It could take many decades to clear the world's cities of slums. From what you know already, suggest one reason why it will be a slow process.

A

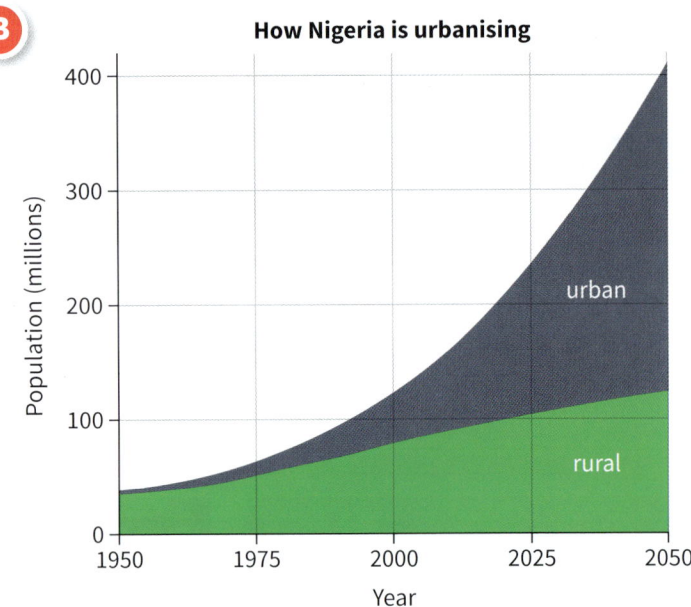

B How Nigeria is urbanising

5 Page 33 shows Amsterdam, the capital of the Netherlands.
   a Define the term *sustainable city*.
   b Explain how this is helping Amsterdam toward sustainability:
      i 40% of all travel in the city is by bike.
      ii The city has a large efficient power plant where household waste is burned to make electricity.
   c Amsterdam has many canals. (The photo shows some.) Suggest a way it could use these to improve sustainability.

6 *Urbanisation is a harmful process for the human race.*
   To what extent do you agree with the statement in italics? Decide, then justify your decision. Write at least half a page.

50

4 Coasts

A ▲ Stoer, Scotland.

C ▼ Sennen Beach, Cornwall.

B ▲ Morston Marshes, Norfolk.

D ▼ Handfast Point, Dorset.

4.1 What causes waves and tides?

 At the seaside you'll notice the waves, and the tides. What causes them? Find out here.

> **Did you know?**
> - The prevailing (most common) wind in the UK blows in from the south west.

What causes waves?

Waves are caused by the **wind** dragging on the surface of the water. The length of water the wind blows over is called its **fetch**.

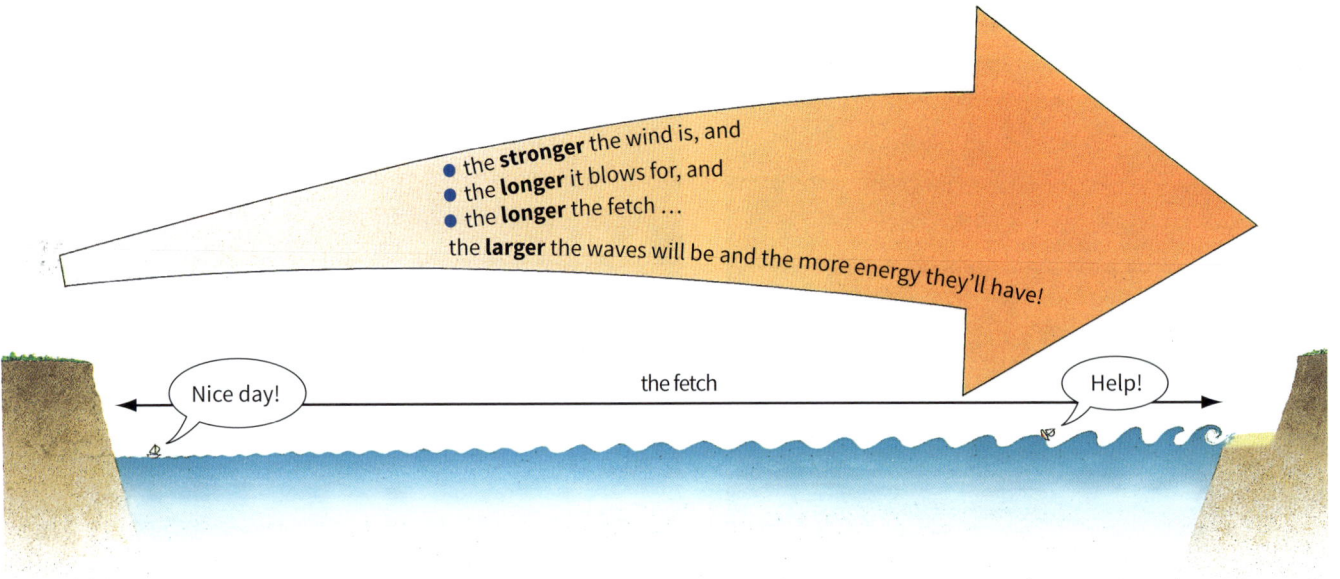

- the **stronger** the wind is, and
- the **longer** it blows for, and
- the **longer** the fetch …

the **larger** the waves will be and the more energy they'll have!

Nice day!   the fetch   Help!

When waves reach the coast

Out at sea, the waves roll like this. In a gale they can be over 30 m high! But at the shore they break …

… giving turbulent water called **swash**. The water rushing up the sand is called the **uprush**.

The water rolling back into the sea is called the **backwash**. Another wave will arrive shortly.

If the backwash has more energy than the uprush, the waves eat at the land, dragging pebbles and sand away. (This happens with high steep waves.)

But if the uprush has more energy than the backwash, material is carried onto the land and left there. (This happens with low flat waves.) The material builds up to make a **beach**.

Tides

Even when the sea is calm and flat, the water level is always changing. That's mainly because of the **moon**. As the moon travels around Earth, it attracts the sea and pulls it upwards. The **Sun** helps too – but it is much further away, so its pull is not so strong.

The rise and fall of the sea gives us the **tides**. Look at these photos:

> **Did you know?**
> - The Bristol Channel has the 2nd largest tidal range in the world …
> - … of up to 14 metres.

Thursday 17 May, at noon. The tide is **out**, here at Porthcawl in Wales. In fact the sea has reached its lowest level for today. This is called **low tide**.

Same place, same day, 6.15 pm. Now the tide is **in**. The water has covered the beach. The sea has reached its highest level for the day. This is called **high tide**.

High tides occur about every twelve and a half hours, with low tides in between. The drop in sea level from high to low tide is called the **tidal range**. It changes through the year, because the pull on the sea changes as the moon orbits Earth, and Earth orbits the Sun.

Your turn

1. Three factors determine the height of the waves in a place. Name them.

2. The arrows below are winds blowing onto island **X**.

   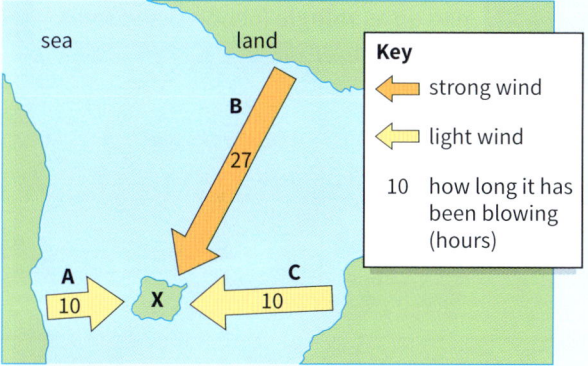

   Which wind will produce:
   a. the largest waves
   b. the smallest waves
   at the coast of X? Explain your answers.

3. Now think about the waves around your own island.
   a. The *prevailing wind* in the UK is a *south west wind*. Define the terms in italics. (Glossary?)
   b. Explain why the south west tip of England gets some really high waves. (Check pages 140 – 141.)
   c. Most of the UK's surfing schools are in south west England, and Wales. Why?

4. Using full sentences, explain what these terms mean:
   a. swash   b. uprush   c. backwash

5. This question is about the four photos on page 51. Each time, give evidence to support your answer.
   a. Which is stronger, the uprush or the backwash:
      i. in photo **C**?   ii. in photo **A**?
   b. Which photo shows the coast at an *estuary*?
   c. In which place do the waves have least energy?

6. a. What are *tides*, and why do they occur?
   b. Photo **B** on page 51 was taken at low tide. What difference(s) would you expect to see at high tide?
   c. Now repeat **b** for photo **C**.

7. Look again at photo **A** on page 51. Yesterday you were exploring the rocks – and got trapped at **X** by the high tide! Write a blog about the event, and how you escaped, or were saved.

4.2 What work do the waves do?

Here you will learn how waves shape the coastline.

The waves at work

Waves work non-stop, night and day, year after year, shaping the coastline. This shows what they do.

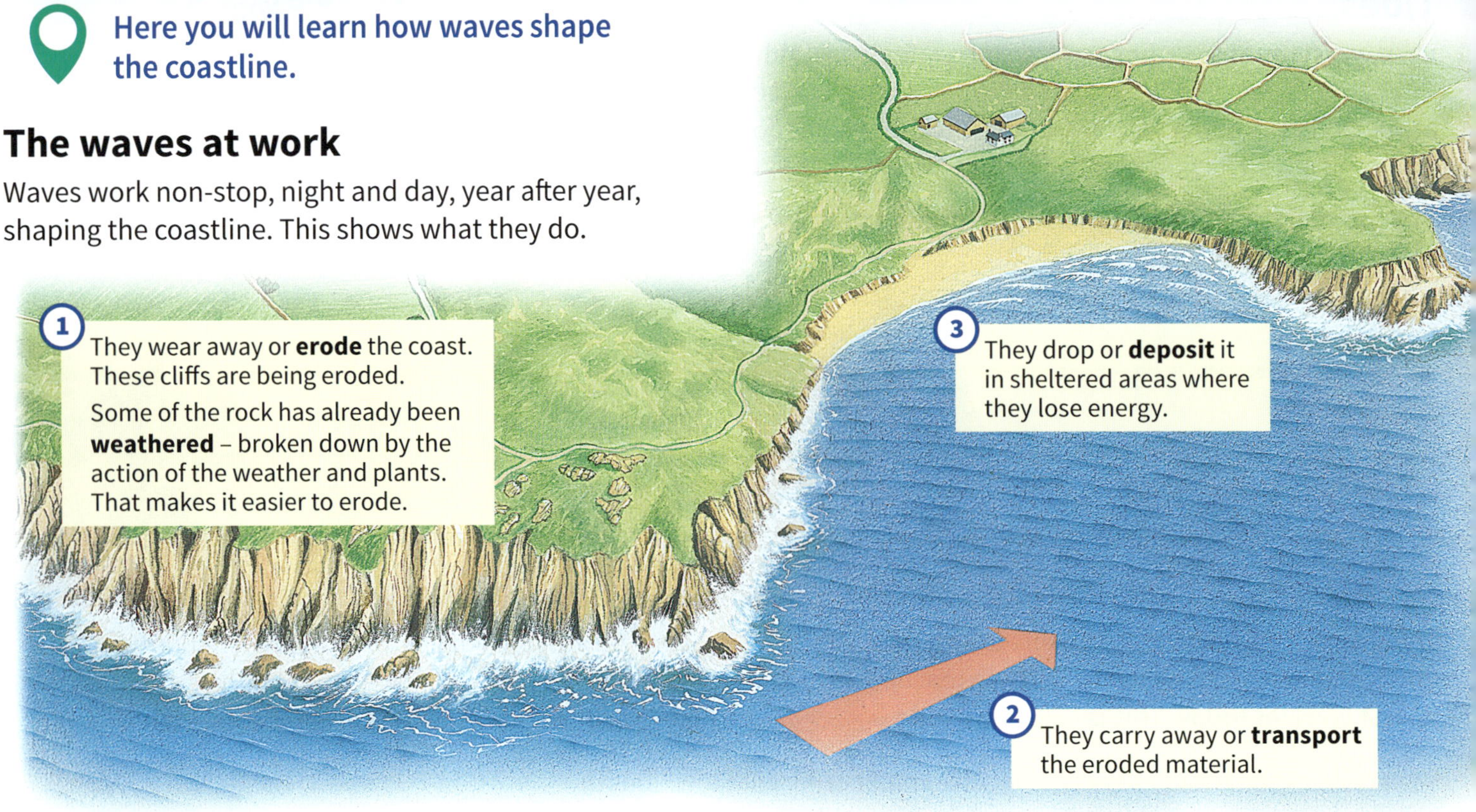

1. They wear away or **erode** the coast. These cliffs are being eroded. Some of the rock has already been **weathered** – broken down by the action of the weather and plants. That makes it easier to erode.

2. They carry away or **transport** the eroded material.

3. They drop or **deposit** it in sheltered areas where they lose energy.

Now we will look at each of these in more detail.

Erosion

This is how waves wear away the coast:

- They pound at the rock like a hammer. Over time, this breaks the rock up.
- They force water into cracks in the rock. That helps to break it up. It's called **hydraulic action**.
- They dissolve soluble material from the rock. This is called **solution**.
- They fling sand and pebbles against the rock. These wear it away, like sandpaper. This is called **abrasion**.
- The bits of rock that break off also get worn down, by knocking against each other. This is called **attrition**. They get smaller and smaller. They end up as **shingle** (pebbles) and **sand**.

The more energy the waves have, and the softer the rock, the faster erosion will be.

Coasts

Transport

The waves carry the eroded material away. Some is carried right out to sea. But a lot is carried along the coastline. Like this …

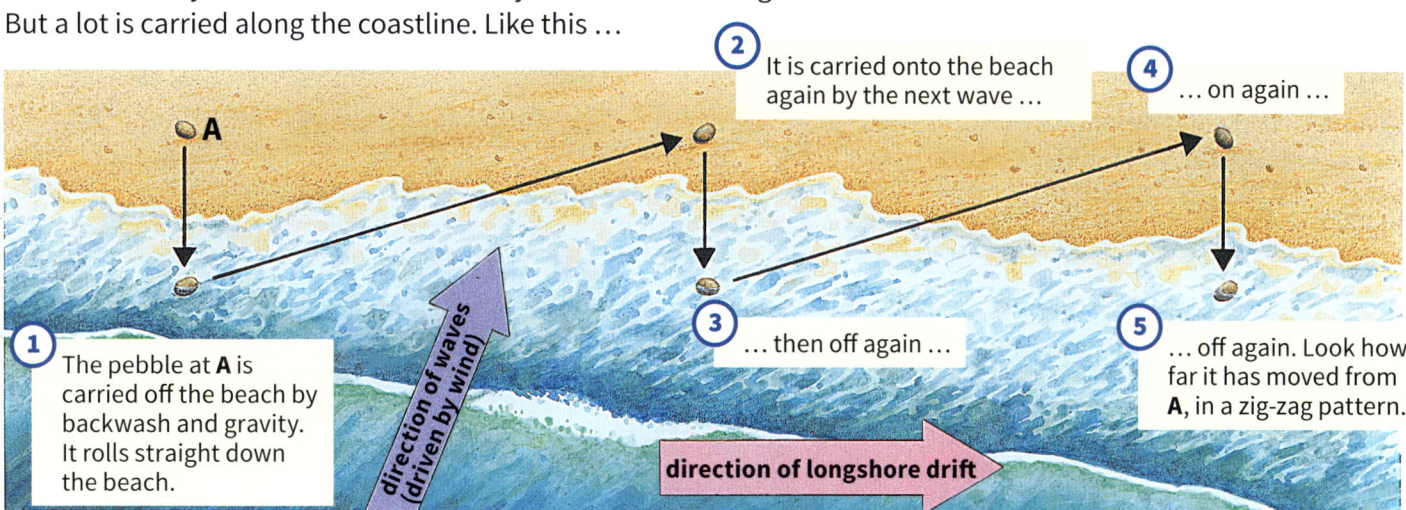

1. The pebble at **A** is carried off the beach by backwash and gravity. It rolls straight down the beach.
2. It is carried onto the beach again by the next wave …
3. … then off again …
4. … on again …
5. … off again. Look how far it has moved from **A**, in a zig-zag pattern.

direction of waves (driven by wind)

direction of longshore drift

Hundreds of thousands of tonnes of pebbles and sand get moved along our coastline every year, in this way. The process is called **longshore drift**.

Many seaside towns build **groynes** of wood or concrete, to stop their beaches being carried away by longshore drift. Look at this photo.

Deposition

Waves continually carry material on and off the land. If they carry more *on* than *off* – a **beach** forms!

Beaches form in sheltered areas, where the waves have less energy. Low flat waves carry material up the beach and leave it there. Some beaches are made of sand, and some are shingle (small pebbles).

▲ Groynes stop the beach being carried away.

Your turn

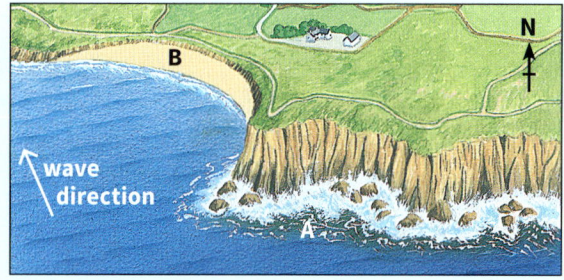

1. Waves do three jobs that shape the coastline. List them.
2. Describe three ways in which waves erode rock.
3. X and Y are beach pebbles of the same type of rock.
   a. Which one has been in the water for longer? Explain.
   b. Name the process that made Y so smooth.

4. Look at the groynes in the photo on the right above.
   a. Define the term *groyne*. (Glossary?)
   b. Are those groynes working? Justify your answer.
   c. In which direction is the longshore drift heading?
      i. towards the north east
      ii. towards the south west
      Give your evidence. (The N arrow will help.)
5. The prevailing wind is blowing, in the drawing above.
   a. What direction is it blowing from? (Look at the waves!)
   b. There is a beach at B. Give a reason.
   c. Where might some of the sand at B have come from?
6. Does all the rock along Britain's coast erode at the same rate? Decide, and give your evidence. (Page 139?)
7. *If you have studied rivers*, write a paragraph comparing the work that rivers and the sea do, in shaping land. Identify similarities and differences.

4.3 Which landforms do the waves create?

This is about the landforms which the waves create along the coast, by eroding and depositing material.

Sculptor at work

This coast is made of different rocks, some hard, some soft. It was once straighter. But look at it now!

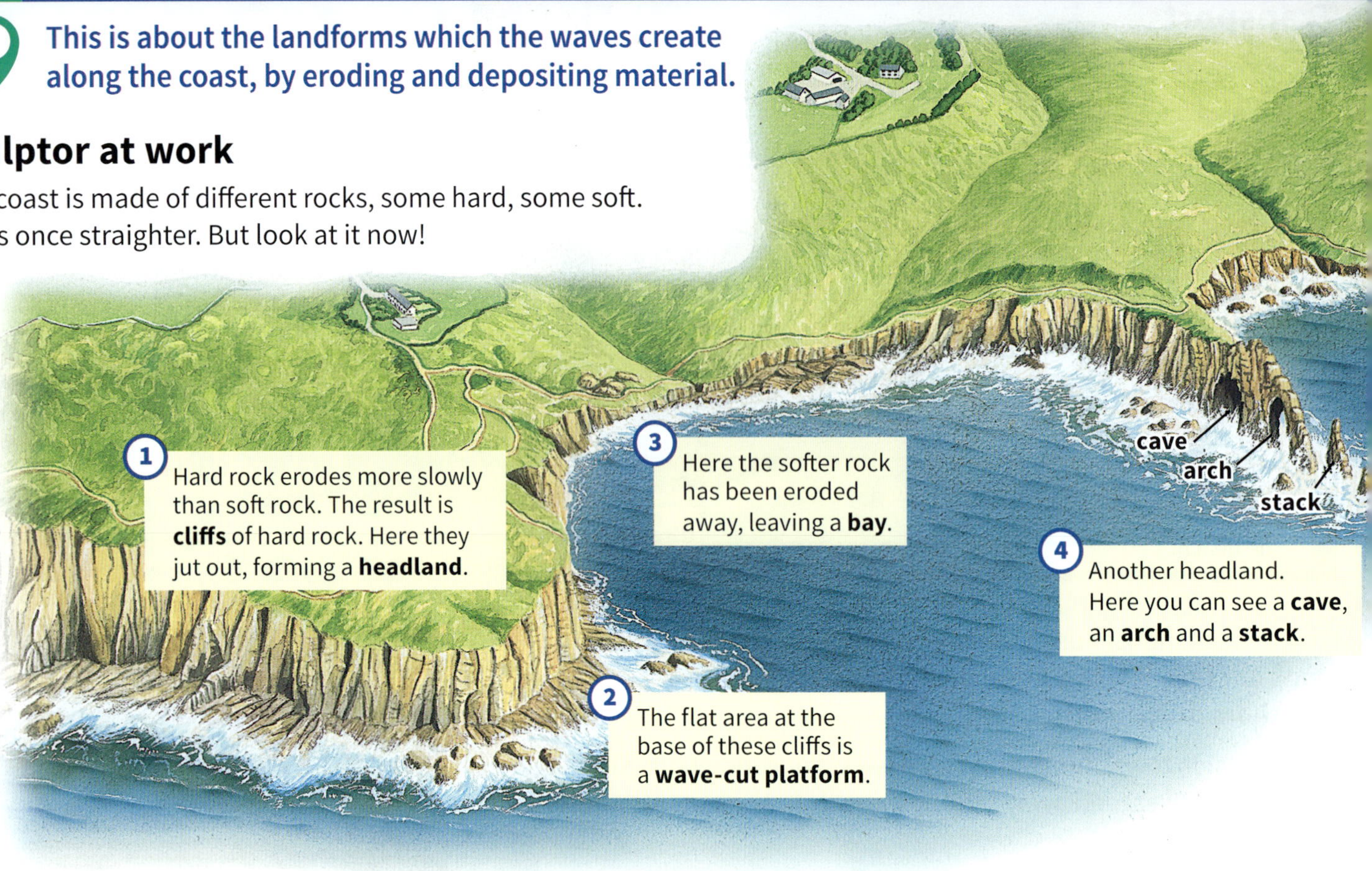

1. Hard rock erodes more slowly than soft rock. The result is **cliffs** of hard rock. Here they jut out, forming a **headland**.
2. The flat area at the base of these cliffs is a **wave-cut platform**.
3. Here the softer rock has been eroded away, leaving a **bay**.
4. Another headland. Here you can see a **cave**, an **arch** and a **stack**.

How a wave-cut platform forms

The waves carve **wave-cut notches** into cliffs at a headland. These get deeper and deeper …

… until, one day, the rock above them collapses. The sea carries the debris away.

The process continues non-stop. Slowly the cliffs retreat, leaving a **wave-cut platform** behind.

How caves, arches, and stacks form

The sea attacks cracks in the cliff at a headland. The cracks grow larger – and form a **cave**.

The cave gets eroded all the way through. It turns into an **arch**. Then one day …

… the arch collapses, leaving a **stack**. In time, the waves erode the stack to a **stump**.

Coasts

6 Some is deposited in sheltered areas like this bay, forming a **beach**.

5 Eroded material is carried along the coast by longshore drift.

7 Here's another headland. But the longshore drift continues in its usual direction …

10 Silt and mud may build up in this sheltered area behind the spit. It becomes a **salt marsh**.

9 The end of the spit is curved by the waves.

8 … so sand and shingle are deposited here, in the sea. They build up a **spit**.

Your turn

1.

Landform	Definition	Created by … erosion	deposition
cliff		✔	
headland			

Make a table like the one started here. Use the width of your page, with a wider column for the definitions.

Complete the first two columns for all the terms in bold in this unit. (Glossary?) Then put a ✔ in the third or fourth column to show how each landform was formed. (One ✔ has been put in for you.)

2. Draw a sketch of the landscape in photo **A**.
   a. On your sketch, label:
      a wave-cut notch   a wave-cut platform   an arch   a stump
   b. Explain how the arch was formed.
   c. Draw dotted lines to show where there was once another arch.
   d. What will happen to the stump over time?

3. Over time, vegetation may grow on spits. Photo **B** shows the spit at Dawlish Warren in Devon.
   a. Draw a labelled sketch of the spit. Add the N arrow, an arrow showing the direction of longshore drift, and a title.
   b. Add notes to show where humans have interacted with the spit. (Look closely!)

4. Make a larger copy of the drawing below to show how the coastline might look 10 000 years from now. Label any landforms, and add notes to explain how they formed.

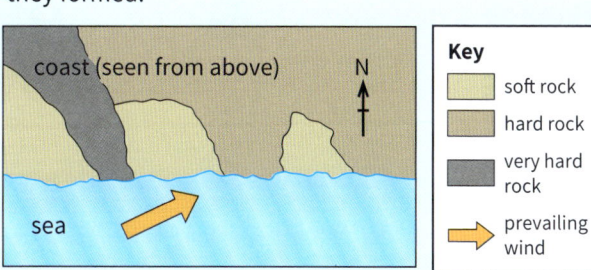

Key:
- soft rock
- hard rock
- very hard rock
- prevailing wind

A

B

4.4 What do we use the coast for?

We use the coast in many different ways. You can find out about them here.

The coast and us

There is coast all around us, in Britain. A few million of us live at the coast. And we all live within 113 km of it. So what do we use it for? Look!

C We light up your life.

A It's great to get away.

B This used to be a huge industry in the UK. It has shrunk!

D Just in from China.

E Baaa.

F We need to be in a place like this, where we can ship in raw materials.

G We all love the sea view and fresh air.

H We'll defend you.

I We take sand and gravel from the sea bed, off the coast. Millions of tonnes a year … … for things like road building.

Coasts

Conflict over land use

As the photos show, the coast is used in different ways. And this often leads to **conflict**!

For example people may not want factories in scenic places. Some object to caravan parks. To reduce or avoid conflicts, places need to be **managed**. The local council plays a big part in this.

We'll explore some conflicts in *Your turn*.

Changes in land use

Table **J** shows land use at the coast.

(A hectare is about the size of one-and-a-half football pitches.)

Like land use everywhere, the land use at the coast changes over time.

For better or for worse?

See what you think.

Did you know?
- Around 55% of the land at the coast belongs to the state.
- Around 20% is in private hands.
- Local councils own most of the rest.

J Changes in land use at the coast (England, Wales, N Ireland)

Land use	Area in hectares		% change
	1965	2015	
Open countryside (including farmland)	353 980	339 180	– 4 %
Urban (built-up)	42 070	59 630	42 %
Woodland	20 900	29 280	40 %
Leisure and sporting activities	14 630	17 990	23 %
Industry	9430	13 080	39 %
Caravan sites	4870	7020	44 %

Your turn

1. Look at the photos on page 58. They show ways we use the land at the coast, and the sea off it.
   You have to match each photo to a term in the box below. But the terms are all jumbled. So sort them out first. (Start with the easy ones?)

ettlessment	surelie	fragnim
dustriny	gredding	hisginf
dwin rowpe	trops	feendec

   Give your answer like this: *A* =

2. Some of the ways the coast is used benefit you, even if you do not live there.
   From your answers to **1**, identify the coastal land use / activities which benefit you. Justify each choice you make.

3. a Most leisure activities at the coast are at **seaside resorts**. Name five seaside resorts in the UK. (Page 139 may help.)
   b Now name three British cities with ports.

4. a Land use can cause conflict. Define *conflict*.
   b Explain why there might be conflict over the land use / activities shown in this pair of photos:
      i   B and I   ii   F and G   iii   F and A
   c Choose another pair of photos where the contrasting land use /activities might give rise to conflict.
     Explain why you chose those photos.

5. Table **J** shows how coastal land use changed over 50 years, in England, Wales and Northern Ireland.
   a i   What is the most common land use at the coast?
     ii There was less of this type of land in 2015 than in 1965. What % less?
   b Which land use showed the biggest % increase?
   c i   How did the amount of woodland change?
     ii Do you think this change was for the better, or worse? Decide, and justify your answer.
   d Repeat question **c**, but this time for industry.
   e The number of homes at the coast increased across the 50 years. How can you tell, from the table?

6. Look again at the land use for 2015, in table **J**.
   a How would you display this data, to make it easy to compare the amounts of land? Discuss with a partner.
   b Then display the data for 2015 in the way you think best.

7. In 1965, university students walked along the coast for weeks, to measure the land use in table **J**. But in 2015 it was measured from a desk. Explain how this was possible.

8. The UK is wealthy compared to most countries. Outline ways in which its coast has helped to make it wealthy.

9. What are the *disadvantages* of having a coastline? Identify as many as you can. Answer in any way you choose. (Spider map? drawings? bullet points?)

4.5 Your holiday in Newquay

In this unit you'll find out about Newquay, in Cornwall – with the help of an OS map.

Did you know?
- So far, the highest wave ever surfed was 24.4 m …
- … off the coast of Portugal.

Meet Newquay

The OS map on the next page shows Newquay, on the coast of Cornwall. 400 years ago, it was just a small fishing village. Now it's a seaside resort – and a surfers' paradise.

Newquay has around 22 000 full-time residents – and about five times as many people in the summer. You are about to join them for a holiday.

Your turn

1. What can you tell about the coast at Newquay, from page 61? Is it flat? Smooth? Sandy? Describe it in four sentences.
2. Find Towan Head on the map. (It's in square 7962.)
   a. What is *Head* short for?
   b. How was Towan Head formed?
   c. Find two other examples of this landform on the map to the east of Towan Head. Give their names.
3. *Newquay has different types of rock.*
   Give evidence from page 61 to support this statement.
4. Look at the tiny island marked X in photo **B**, with a footbridge. It's at 811618 on the OS map, near the aquarium.
   Do you think the island was always like this? Explain.
5. **C** is an aerial photo of Newquay. Places i – v are marked on it. Name them, with help from the OS map. (Four are beaches.)
6. Now, holiday time. You are off to Newquay on a five-day camping holiday, in August, with your friends.

I'm hungry. / I'm tired. / Hurry up, you lot.

   a. They let you pick a camp site from the map. Which one will you choose? Give a 6-figure grid reference for it, and say why you chose it. (There's an OS key on page 138.)
   b. The map gives plenty of clues about things to see and do in Newquay. List as many as you can. Then tick the ones that appeal to you.
   c. You will go there by train. Where is the station? Give a 6-figure grid reference for it.
   d. From the station, you'll go straight to the tourist information office, to find out about surfing lessons.
      i. About how far is the office from the station?
      ii. In which direction will you walk?
7. You book some surfing lessons. They will be at Lusty Glaze beach. (See photo **A**.)
   a. You will walk to lessons from your camp site.
      i. Draw a sketch map of your route, with any landmarks.
      ii. Calculate how long the route is.
   b. Lessons will start at 9 in the morning. By what time do you think you'll need to leave the camp site?
8. You see lots of other surfers, from the UK and many other countries. Some flew in to Newquay Cornwall airport.
   a. Find the airport on the map. Give a grid reference for one square of it.
   b. The airport had around 457 000 passengers in 2018. It hopes to attract more flights and at least 600 000 passengers a year by 2030. (People flying in won't all stay in Newquay. The airport serves all of Cornwall.)
      i. How might more flights benefit Newquay? List as many benefits as you can.
      ii. What problems might they cause?
9. A change is **sustainable** if it continues to benefit us *economically* and *socially*, into the future, *without harming the environment*.
   a. Discuss whether an increase in flights into Newquay Cornwall airport could be sustainable. (Think about each term in italics. Glossary?)
   b. i. Imagine you are in charge, in Newquay. On balance, will you allow more flights? Explain your decision.
      ii. If your answer to **i** is *yes*, you might want to set some conditions. For example about noise, or pollution. Suggest three conditions you might set.
10. Home again. Now write a blog about Newquay for your website. Say where it is, what it's like, and what you did there. What photos will you add? And what about a map?

4.6 Storm surge!

Rivers can flood streets, and homes. So can the sea! Find out more here.

It's the boss!

It's wonderful being surrounded by sea. It means beaches, and water sports, and sea views, and fun. But every so often, the sea reminds us that it is the boss. We cannot control it.

Storm surge, December 2013

5 and 6 December 2013. Some people will never forget those dates, when a savage sea battered our coast. It was the worst storm in 60 years.

Gale-force winds piled big waves onto the coast. Seawater flooded over the sea walls built to keep it out. It swept along streets, and into shops and homes.

Britain's east coast got the worst of it. A high volume of water rolled down the North Sea, in a storm surge. The waves lashed the coast. Thousands of people were evacuated from their homes, for safety. Hundreds of homes were flooded.

In Norfolk, the raging sea ate chunks from the soft cliffs. In the village of Hemsby, the cliff fell away below seven homes. Three slid into the water, and were carried off like toys.

The Thames Barrier had to be closed, to protect London.

But there was some comfort too. Coastal defences in many places held up well, protecting hundreds of thousands of properties. Without them, the damage wreaked by the sea would have been much much greater.

▲ The sea flooded roads and homes in Rhyl, north Wales, on 5 December 2013.

▲ In Hemsby, Norfolk, the sea ate into these soft cliffs.

▲ Prestonpans in Scotland, under attack.

▲ The sea flooded into Scarborough.

Coasts

Why did it happen?

Three factors combined to create the storm that battered the coast.

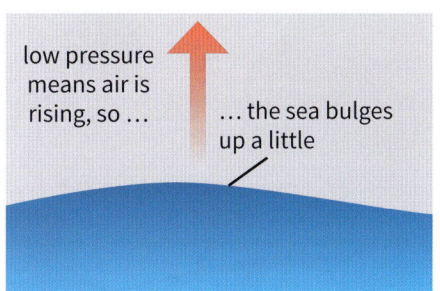
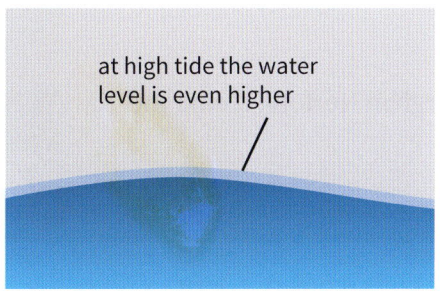
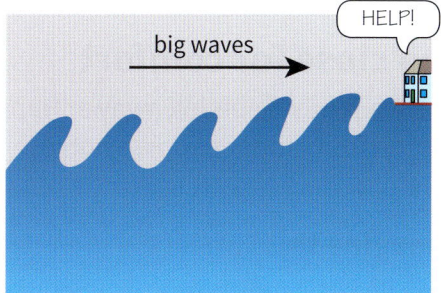

Air pressure was low. That means air was rising. When this happens, the sea is drawn upwards a little.

The **high tides** were also at their highest, at that time of the month. This added to the water level.

But the main factor was the **strong winds**. They whipped up the waves and pushed them onto the coast.

The storm surge at the east coast

The east coast suffered most. Map **E** shows why.

1. The system of low pressure passed north of Scotland, heading east.
2. The winds and high water level caused the **storm surge** – a big surge of water. It swept down the North Sea. The waves, whipped by a north wind, hammered the east coast.
3. The surge grew higher as it moved further south, where the North Sea gets narrower and more shallow. So the flood risk grew too.
4. Then the surge headed north again, flooding the coasts of other European countries.

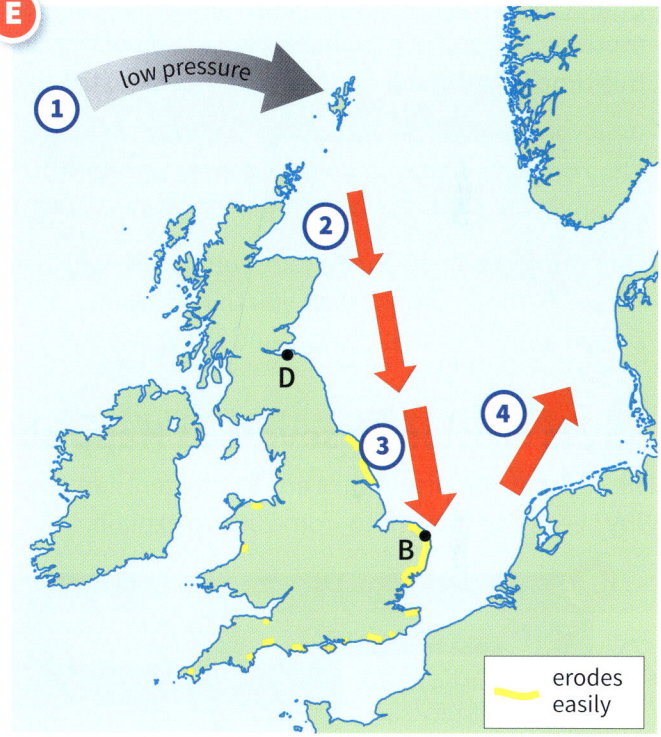

Now look at the yellow areas on the map, along Britain's coast. They show where the rock is soft or quite soft, and easily eroded.

As you can see, several parts of the east coast are easily eroded. So they are at risk – and especially in a storm surge. B marks Hemsby where photo **B** was taken.

Your turn

1. Look at photo **D**. The waves are lashing the promenade. State the *main* cause of these big waves.
2. The factors below played a part in the scenes in photos **A – D**. Explain how they contributed. Write full sentences.
   a. There was low air pressure over the UK.
   b. The high tides were at their highest.
3. There was a storm surge in the North Sea.
   a. Define the term *storm surge*. (Glossary?)
   b. Check B and D on map **E**, and their photos. Explain why:
      i. the surge was higher at B than at D
      ii. the damage was more permanent at B
4. Some places were saved by *coastal defences*. What are these?
5. The Thames Barrier lies on the Thames to the east of London. (See the photo on page 67.) It is closed during storm surges. Suggest reasons why protecting London has high priority.
6. Look at photo **B**. That's your house on the right!
   a. Write a letter to the sea, describing how this storm surge has affected you.
   b. To your great surprise, you get a reply from the sea, written on waterproof paper. What does it say?
7. Could storm surges become more frequent? Discuss!

4.7 How long can Happisburgh hang on?

This is about erosion in one village on the UK's east coast – Happisburgh in Norfolk. (Say *Haze-boro*.)

Too close for comfort

The sea is busy nibbling away at Happisburgh in Norfolk. It eats the ground from under people's homes.

Imagine this …

You live in Happisburgh, in a house on the cliffs. Look out your window and there's the sea. You are minutes from the cliff edge.

And that's the problem. You can't relax. The sea makes you nervous.

At night when the wind is strong, you hear the waves roar. From time to time there's a crash, and the whole house shakes. In the morning you run out to check. How much of the cliff has collapsed this time? How much closer is your house to the sea?

When you were little, lots of houses stood between you and the sea. They've gone. Some slid over the edge during storms. The rest were demolished when the sea got too close, for safety.

One thing you know for certain: your house will disappear too. But when? Perhaps sooner than you think. Because climate change is bringing more severe storms, and rising sea levels …

A

▲ Going, going, … The cliff collapsed under this house on Beach Road in the storm surge of 2013. The house is marked ⬤ in photos B and C.

Why is erosion so severe at Happisburgh?

The coast at Happisburgh is eroding faster than anywhere else in the UK. Why? Let's use this photo from 2009 to explain.

B

① The main reason is that the cliffs are soft – sand on top and clay below.

② Rain soaks into the cliffs and helps to weaken them. The more it rains, the weaker they get.

③ Meanwhile, the sea erodes them from below. Not so much in summer, when the weather is calmer …

④ … but winter brings strong north winds, and storms, and sometimes storm surges. Waves lash the cliffs, and chunks collapse.

⑤ Because the cliffs face north east, they are even more vulnerable to winter storms.

⑥ **Groynes** help to slow down erosion. (The sand they trap absorbs energy from the waves.) But these wooden groynes have broken down.

⑦ These wooden fences or **revetments** are meant to slow down erosion, by making waves break early. But they were damaged in a past storm, and not repaired.

⑧ A line of big rocks, called **rock armour**, has been put in place to protect the cliffs from the full force of the waves. It works – but there isn't enough of it.

Happisburgh, August 2009

Coasts

Can nothing be done?

As you saw in **B**, they have tried different types of defences to protect the cliffs at Happisburgh. These cost a lot – and the government paid. But over time, the sea broke up the wooden ones. Repairs would cost more money.

Now the government says that places like Happisburgh will no longer be protected. They will be left to the sea. Find out more in the next unit.

> **Did you know?**
> - Low-lying coastal areas of the UK are likely to be under water by the end of this century.

Your turn

1. Where in the UK is Happisburgh? On which coast? By which sea? Describe its location as fully as you can. (Page 139?)

2. Look at the house in photo **A**. Explain the part this played in its destruction:
   a. the material the cliffs are made of
   b. rain
   c. the direction the cliffs are facing
   d. the sea

3. a. Photo **C** shows wooden barriers on Happisburgh beach.
      i. Name the barriers at right angles to the cliffs, and explain how they could help to slow down erosion.
      ii. Repeat **i** for the barriers parallel to the cliffs.
   b. Photo **B** shows a barrier that's not wooden. Name it.
   c. Which of the three types of barrier would you expect to last longest? Explain your answer.

4. a. Identify the white shapes towards the top of photo **C**.
   b. What advantage do they have over the houses, given the risk of erosion?

5. Compare photos **C** and **D**, taken from different angles.
   a. Which photo was taken at higher tide?
   b. List all the changes you notice between 1996 and 2019.
   c. In your opinion, have the barriers been effective against erosion? Justify your answer.

6. Look at the church in photo **C**. In January 2014, it stood 141 m from the cliff edge.
   a. It is estimated that the cliffs are eroding at a rate of roughly 3 m a year. Using that figure, calculate in which year the church may be lost to the sea.
   b. But we can't rely on the estimate in **a**, for future erosion. Why not?

7. The government will not pay for more coastal defences at Happisburgh. But the village did get a grant to help it adapt. Some of the money was used to create a new caravan park further inland.
   a. Explain why a new caravan park was needed.
   b. Suggest reasons why a caravan park is important to Happisburgh.

8. You have a big fieldwork project! To measure the cliff loss at Happisburgh over the next 12 months.

   Work with a partner to make a plan. Then write down the steps you will follow, and draw a sketch to show your method.

   (Think about what to measure, and where, how, and how often. How will you make sure you measure from the same place(s) each time? How will you stay safe?)

C Happisburgh, November 1996 (Note the circled house.)

D Happisburgh, January 2019 (The same house is circled.)

4.8 How can we protect places from the sea?

 How can we stop places being flooded and eroded by the sea? Find out in this unit.

The threat from the sea

People who live on the coast are at risk from the sea in two ways:
- The sea can flood their places. *So they try to keep it out*. Look at **1** below.
- The waves can erode the land away. *So they try to reduce the waves' energy*. Look at the five methods in **2**. You met some of them already.

Typical costs of coastal defences per metre	
Sea wall	£6000
Rock armour	£3000
Artificial reef	£2000
Wooden revetment	£500
Wooden groyne	£1000
Beach nourishment	£3000

1 Keep the sea out

Sea walls are the usual way to keep the sea out. They are often curved, to reflect the waves away.

2 Reduce the waves' energy

Rock armour (big rocks) soaks up the waves' energy. So it slows the erosion of cliffs and sea walls.

An **artificial reef** made of rock can be built out at sea, so that the waves break earlier, away from the beach.

Revetments are a bit like fences. The waves batter them instead of the cliffs.

Groynes help, because they stop sand being carried away. Sand absorbs some of the waves' energy.

Adding more sand or shingle to a beach helps too. This is called **beach nourishment**.

But the trouble is ...

- Coastal defences cost a lot. Look at the table at the top of this page.
- Most defences break down eventually, or get washed away.
- Defences may do harm elsewhere. For example protecting a cliff from erosion may starve a beach of sand, further down the coast.
- And the biggest challenge ... **climate change**. Earth is warming.
  - As the sea warms the water expands. So sea levels are rising.
  - As ice sheets melt, sea levels will rise further.
  - Storms are also becoming more severe.

  So existing defences may fail.

▶ Scarborough, on the east coast, got new coastal defences in 2005. Cost: £53 million. Now look at photo C on page 62.

So who decides?

The money for coastal defences comes from the government. It has a limited amount of money, so has to make choices.

The plans for the coast

Shoreline management plans have been drawn up for Britain's coastline – to cover the next 100 years. The plans take climate change into account. These are the key points:

1. We'll protect places which have enough homes and businesses to make them worth protecting.
2. We'll also try to protect special places – such as key historical sites, or nature reserves.
3. We will not protect other places! And in time, the sea will drown them. But we will help local people to prepare for this.

▲ The Thames Barrier, closed. It's always closed during a storm surge in the North Sea, to protect London.

Your turn

1. Look at the different coastal defences on page 66.
   a. Which type is designed to stop places flooding?
   b. i. List the defences that reduce the waves' energy.
      ii. Explain *why* reducing the energy of the waves helps to reduce coastal erosion.
   c. Identify the three types of defence that are the most likely to be damaged, or washed away, in a storm.
   d. Which type may need redoing the most often? Explain.
   e. Which type is the most expensive?
2. Would sea walls work for Happisburgh? Give reasons.
3. Look at opinions **a – e** on the right. Choose three, and write replies. You can agree or disagree with the speakers. Give your reasons each time.
4. Using the key points from the shoreline management plans (in the yellow panel above) explain why:
   a. Happisburgh will get no more money for coastal defences
   b. it's likely that London will always be defended
5. a. Give two reasons why climate change is helping to speed up coastal erosion.
   b. Suppose that climate change is faster and more extreme than expected. How might plans for defending the coast change? Make your predictions!

4 Coasts

How much have you learned about coasts? Let's see.

check ✓

1 Photo **A** was taken in Scotland's Orkney Islands. (Page 139.) The rocks are sandstone, around 400 million years old.
   a Powerful waves often lash at these rocks. Where do waves get their energy from?
   b Identify the features labelled P, Q, R and S. (R is a s____.).
   c P was not always like this. Draw a sequence of three labelled sketches to show how P formed.
   d Name one process of erosion that helped to form P.
   e Explain the link between features P and R.
   f Is the ancient sandstone in **A** hard to erode, or easy? Decide, and justify your answer.
   g Features like P – S are *not* found on the coast of south east England, around Happisburgh (Unit 4.7). Explain why.
   h There are no plans for coastal defences in the area in **A**.
      i Define the term *coastal defences*.
      ii Suggest a reason why they are not needed here.

2 Photo **B** shows the seaside resort of Hornsea in East Yorkshire. Its population is over 8000. Find it on page 139.
   a i Name the coastal defences labelled X, Y and Z.
      ii Which of them is designed to stop seawater flooding into the buildings in the photo?
      iii Explain the purpose of the rocks that form Y.
   b Hornsea has a sandy beach.
      i Describe how sand forms.
      ii The sand is carried down the coast to Hornsea from further north. Name the process in which material is carried along the coast by the waves.
   c Hornsea is at risk of losing sand.
      i What is the evidence for this, in the photo?
      ii Name another type of coastal defence which would compensate for lost sand.

3 **C** shows the change in global sea level from 1880 to 2014.
   a Describe the overall trend.
   b Give one reason to explain this trend.
   c Estimate the change in sea level from 1880 to 2014. (Don't forget to state the unit.)
   d If the trend continues, and storms become more severe, what are the likely impacts on Hornsea in photo **B**?
   e For now, Hornsea will continue to get money for coastal defences – but Happisburgh (in Unit 4.7) will not. Suggest a reason for this difference.

4 *Natural processes and human activity both play a part in creating seaside resorts.*
Discuss the statement in italics. Photo **B** may help.
And see page 137 for *Discuss*.

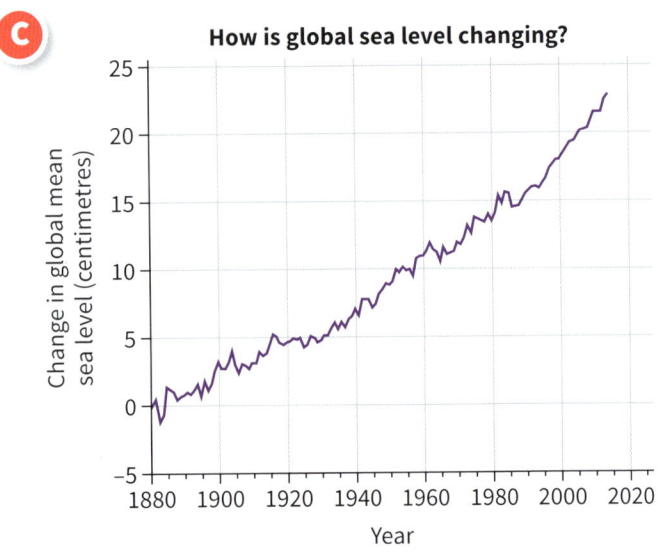

5 Weather and climate

5.1 Weather: what, why, and where?

In this unit you'll look at two big ideas that help to explain all weather.

What's the weather like today?

Weather is the state of the **atmosphere**. What's it like where you are? Warm or cold? Wet or windy? Is it anything like the weather in **A** or **B**?

And why can it be so different in different places?

Why… …do some places have little or no rain?

What if… …there was no rain anywhere?

A

B

What's behind the weather?

Two big ideas help to explain all weather. Let's have a look at them …

1 Earth warms up unevenly, because it's round.

Look at **C**.

- Earth is heated by the beams of energy our Sun sends out as sunlight.
- When they strike Earth's surface they warm it up. It in turn warms the air.
- Because Earth is round, some beams have to warm larger areas than others. So their energy is spread further, and these areas don't get as warm.

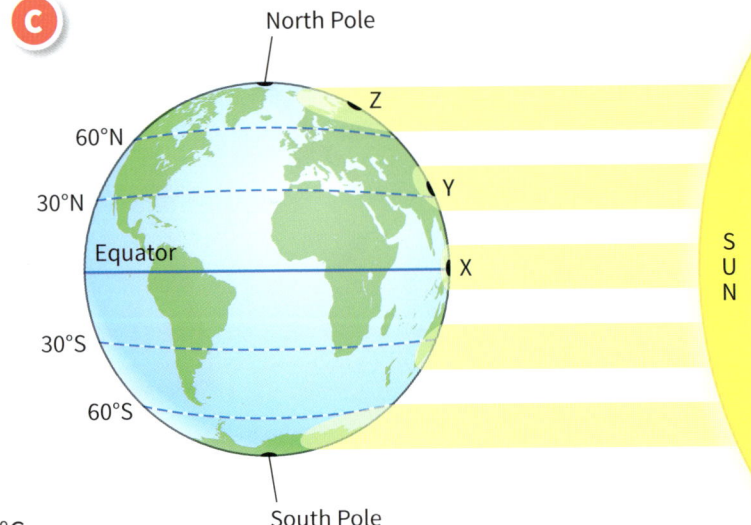

C

Look at the places marked X, Y and Z, for example. It is always colder at Y than at X, because the beam at Y covers a larger area. And Z is the coldest, because that beam covers the largest area.

Now look at the Equator, and the poles. It can be over 30 °C at the Equator when it's −40 °C or less (far below freezing) at the poles. All because Earth is curved!

And that brings us to the second big idea …

70

Weather and climate

2 Then heat gets moved from warmer to colder places.

If heat were not removed from around the Equator, it would get hotter and hotter there. And meanwhile the poles would get colder and colder.

But luckily, heat always moves from warmer to cooler places, to try to even itself out. So warm air moves from the Equator towards the poles. *And this creates weather*. Look:

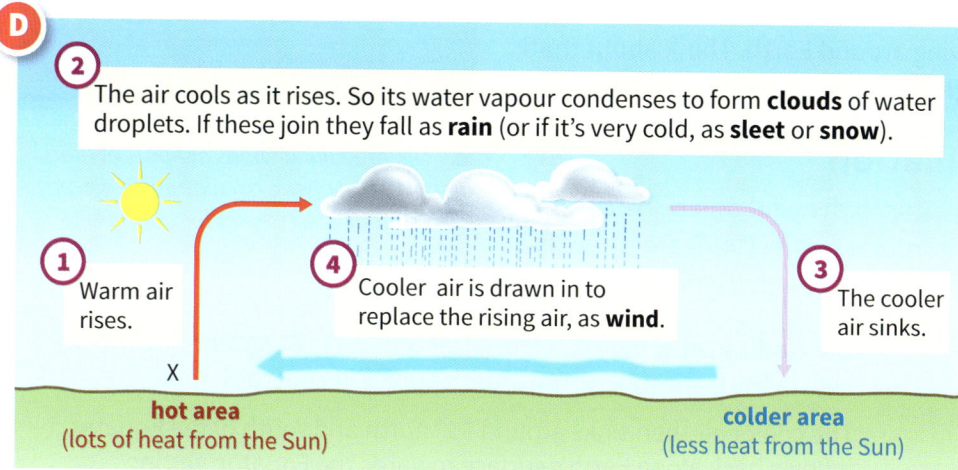

D

1. Warm air rises.
2. The air cools as it rises. So its water vapour condenses to form **clouds** of water droplets. If these join they fall as **rain** (or if it's very cold, as **sleet** or **snow**).
3. The cooler air sinks.
4. Cooler air is drawn in to replace the rising air, as **wind**.

hot area (lots of heat from the Sun) — colder area (less heat from the Sun)

So now we have heat, cold, clouds, rain, and wind. It's the weather!

Where does weather happen?

Look up into the sky. You are looking into the atmosphere. Then look at **F**.

The atmosphere is the layer of gas around Earth. In fact it's a mixture of gases – mainly nitrogen and oxygen, with smaller amounts of carbon dioxide, water vapour and other gases.

Gravity pulls this mixture towards Earth's surface. About 75% of it is in the lowest layer of the atmosphere, called the **troposphere**. This layer is about 13 km deep, on average. It's where you live. You call the gas mixture **air**.

The troposphere is where weather happens. Above it, there's no rain.

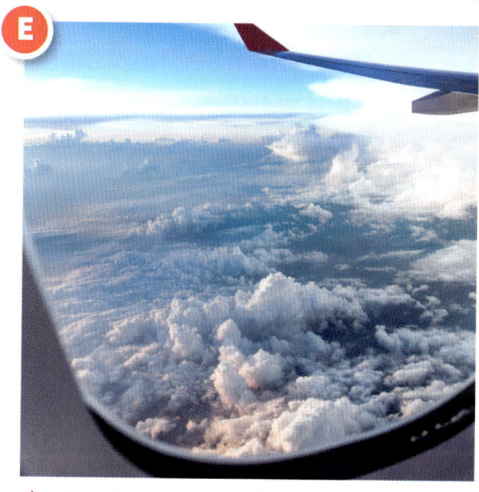

▲ Clouds are made of tiny water droplets. If the droplets grow they will fall as rain.

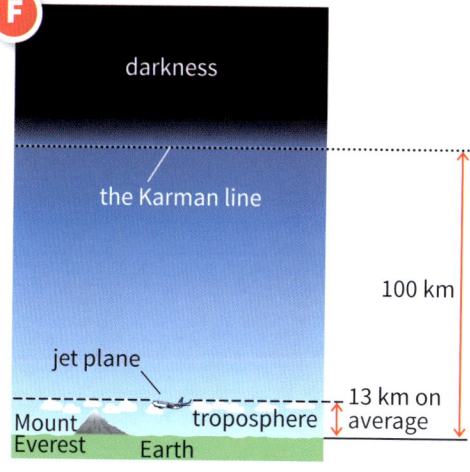

▲ Earth's atmosphere. Scientists argue about where it stops and space begins. The Karman line is often taken as the limit.

Your turn

1. **a** Define *weather*.
   b Write five words (adjectives) to describe the weather:
      i in photo A **ii** in photo B **iii** where you are today

2. Look at diagram C. Explain why:
   a it is warmer at the Equator than at the poles
   b it is cooler in the UK than at the Equator
   You can draw a labelled diagram if that helps.

3. Look at D. The air at X contains water vapour.
   a i What is *water vapour*?
      ii How does it get into air? Explain. (Page 80?)
   b Describe how the rising warm air at X leads to rain.
   c Define *wind*. (Glossary?)
   d Explain how the rising warm air at X leads to wind.

4. Copy this, with the jumbled words unjumbled!
   Our Sun gives out gyeenr as stiglhun. This srawm Earth, which then wrams the ria. Earth is norud so it does not warm leenvy. It's much marrew at the Equator than at the slope. So warm ira then vemos from the Equator to neev things out. This leads to nidw and ainr.

5. How could you demonstrate (in school or at home) that:
   a warm air rises? **b** warm air moves to colder places?

6. You live in the *troposphere*, where weather happens.
   a What is the troposphere?
   b The air is denser in the troposphere than above it. Why?
   c The troposphere is about 13 km deep, on average. To get an idea of its depth, name a place about 13 km from you, on Earth's surface.

5.2 How is heat carried around Earth?

 Here you'll see how air moves around Earth, carrying heat from warmer to cooler places.

Reminder

As you saw in Unit 5.1, Earth is heated unevenly by the Sun, because it's round. So warm air moves continually to cooler places, trying to even things out.

Right now, billions of tonnes of air are moving around Earth. Think about that! This global movement of air is called the **global atmospheric circulation**.

The global atmospheric circulation

Air does not flow straight from the hot Equator to the cold poles. Instead it circulates in bands called **cells**, that curve around Earth. There are three on each side of the Equator. **A** shows a cross-section through them.

▲ The prevailing winds are part of the global atmospheric circulation. Explorers like Captain Cook depended on them to get around.

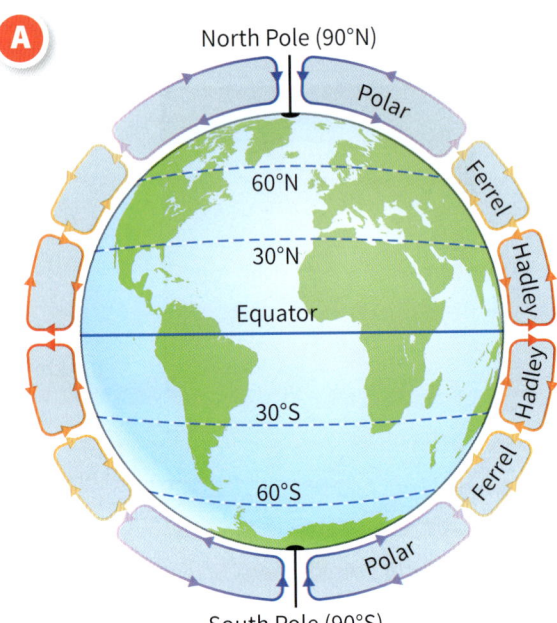

the Hadley cells. Very warm air rises around the Equator, and moves high in the troposphere, towards the poles. It cools as it moves.

The cooler air then sinks around 30° north and south of the Equator. It flows back along Earth's surface towards the Equator, getting warmer … and rises again, carrying more heat away.

the Polar cells. Cold dense air sinks at the poles (90°N and 90°S). As it sinks, it pushes the air at the surface away. This flows towards the Equator, getting warmer.

It rises around 60° north and south of the Equator, and flows back to the poles. There it sinks again, bringing some warmth.

the Ferrel cells. These lie between 30° (where Earth is still very warm) and 60° (where it's cool). They are the 'mixing' cells. Here, warm and cold air get whisked together by **depressions**. (Unit 5.5.) The arrows show the *overall* direction of air flow.

But in real life the cells are not nearly as neat as in **A**! They are affected by the sizes and shapes of the continents, and oceans, and mountain ranges. They also slide up and down a bit, from summer to winter.

The winds at Earth's surface

Look at the arrows at the bottom of the cells in **A**. They represent the air moving along Earth's surface, to complete the loops. They are Earth's **prevailing winds**, and they follow a pattern. Look at **B**.

The winds appear to follow a curved path. That's because Earth is spinning on its axis as they flow.

The winds flowing towards the Equator are turned or **deflected** towards the west. The winds flowing towards the poles are deflected towards the east.

This is called the **Coriolis effect**. It affects *all* the global air flow.

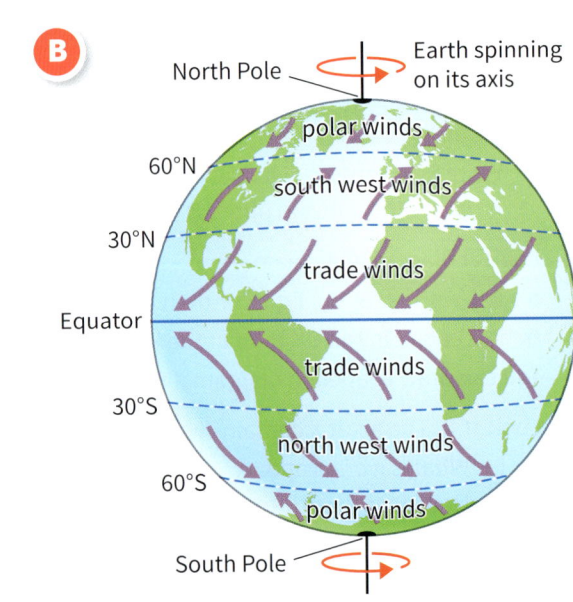

Weather and climate

Belts of high and low pressure

Air pressure is the force pressing down on us, due to the weight of the atmosphere.

- When warm air is *rising* there's less air above us. (It flows away at the top of the troposphere.) So the air pressure falls. We get **low pressure**.
- When cool air is *sinking* there is more air above us. (It flows in at the top of the troposphere.) So the air pressure rises. We get **high pressure**.

As you saw in **A**, air rises in one part of a circulation cell, and sinks in another. The result is belts of high and low pressure around Earth. Look at **C**.

The arrows show the prevailing winds. They blow *from high pressure to low*, but they are deflected by the Coriolis effect.

Why does air pressure matter?

Why does air pressure matter? Because it's closely linked to weather!

When there's high pressure, the sky is clear. No rain. No clouds to block out sunlight. The Sahara is hot and dry because it's in a high pressure belt.

But low pressure means clouds and rain. There's low pressure at the Equator – and it's a very wet region! It usually rains heavily every day.

We'll look more closely at air pressure and weather in the next unit.

Ocean currents

What about the oceans? They cover far more of Earth's surface than land does: 71%.

They too are warmed by the Sun. Like the land, they get warmed most around the Equator.

The result is **ocean currents**: currents of warm water which carry heat to colder areas, and currents of cold water which flow to warmer areas, where they warm up.

Look at **D**. The currents don't flow in straight lines. They are influenced by the shape of the continents, and by the prevailing winds, and the Coriolis effect.

▲ At the Equator in Africa. Low pressure!

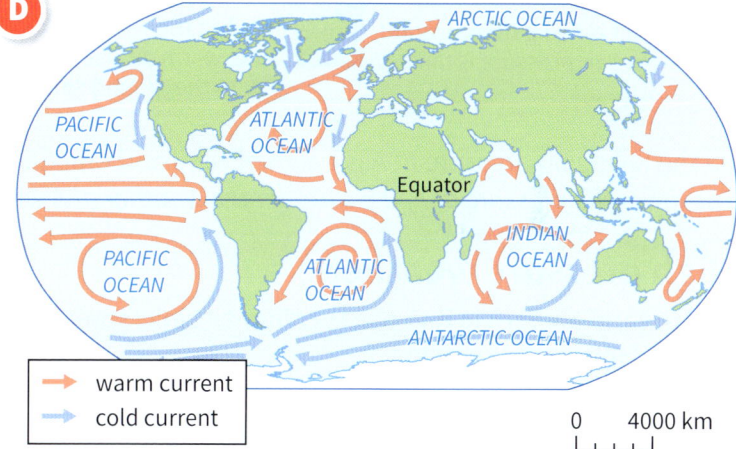

Your turn

1. a What does the global atmospheric circulation do?
   b Name the cells that make up this circulation.
2. Explain how:
   a the Hadley cells stop the Equator getting too hot
   b the Polar cells help to warm the poles
3. The UK lies between 50° and 61°N. Find it on **A**.
   a i Within which cell does the UK lie?
      ii What goes on in that cell?
   b The UK lies near the boundary of another cell. Name it.
4. The circulation cells give Earth its prevailing winds.
   a Define *prevailing wind*. (Glossary?)
   b Which are the prevailing winds for the UK? (See **B**.)
5. a What causes *air pressure*?
   b How does air pressure change when air: i rises? ii sinks?
   c Explain why there's a lot of rain at the Equator, but the Sahara is very dry.
6. a Ocean currents also help heat to circulate. Explain how.
   b Ocean currents don't flow in straight lines. Why not?

73

5.3 Air pressure and our weather

 Air pressure is closely linked to weather. Here you'll learn more about this, with the weather in the UK as example.

What's air pressure?
As you saw in Unit 5.2, **air pressure** is due to the weight of all the air above us. When air is rising, *air pressure falls*. When air is sinking, *air pressure rises*.

Did you know?
- Earth's atmosphere weighs 5 quadrillion tonnes.
- That's 5 000 000 000 000 000 tonnes.

Low pressure weather
Let's see how the weather changes as air pressure falls.

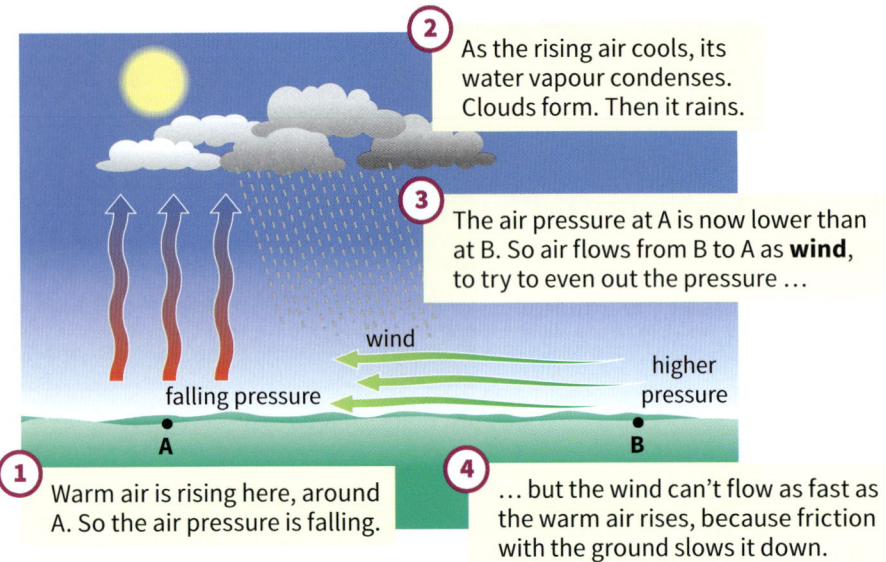

1. Warm air is rising here, around A. So the air pressure is falling.
2. As the rising air cools, its water vapour condenses. Clouds form. Then it rains.
3. The air pressure at A is now lower than at B. So air flows from B to A as **wind**, to try to even out the pressure …
4. … but the wind can't flow as fast as the warm air rises, because friction with the ground slows it down.

▲ Low pressure brings wet and windy weather.

So the falling air pressure at A is a sign of wind and rain on the way. *The lower the pressure at A, the stormier the weather at A will be.*

High pressure weather
When warm air rises in one place, colder air sinks somewhere else – giving high pressure. Look!

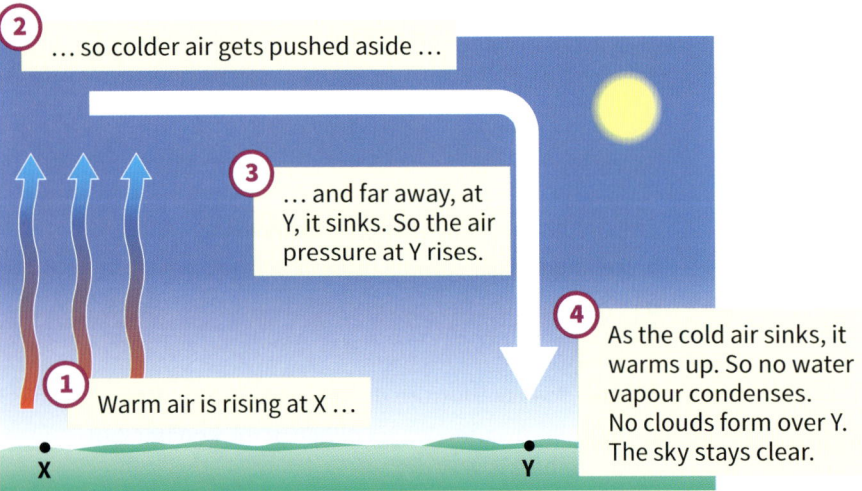

1. Warm air is rising at X …
2. … so colder air gets pushed aside …
3. … and far away, at Y, it sinks. So the air pressure at Y rises.
4. As the cold air sinks, it warms up. So no water vapour condenses. No clouds form over Y. The sky stays clear.

So high pressure means clear cloudless skies. It brings our hottest summer weather and coldest winter weather, as you'll see next.

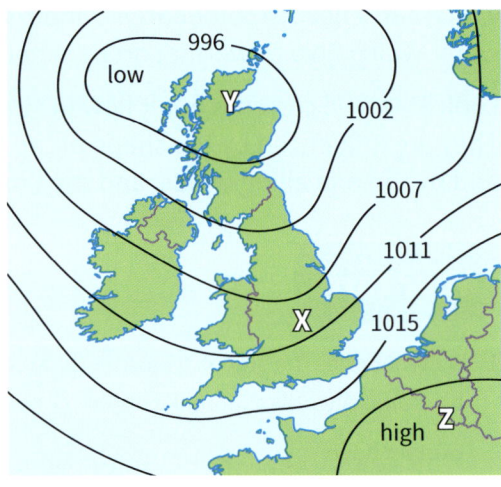

▲ Air pressure is shown using **isobars** – lines which join places at the same pressure. It is measured in **millibars** (mb). Standard air pressure, at sea level, is 1013 mb.

The lower the pressure below standard, the stormier the weather. The higher it is, the more settled the weather.

Weather and climate

When there's high pressure in summer …

There are no clouds in the way, so the sun is strong. Take care you don't get sunburn!

No cloud means no rain. So there may be a hose-pipe ban in some places.

But inland, on very hot cloudless days, the hot air may rise rapidly. It cools, and then tall black clouds form.

Since there is no cloud to trap the heat in, evenings can be cool.

Inside these clouds, strong currents of air whip around, causing **thunderstorms** …

… and thunderstorms can lead to heavy rain, and **flooding**.

No cloud also means the ground gets cold at night. Water vapour condenses on grass to form **dew**.

When there's high pressure in winter …

There is no cloud to act as a blanket. So the days are clear, cold, and bright.

With no cloud, the ground cools fast at night, and cools the air above it. Water vapour condenses and freezes on cold surfaces, giving **frost**.

Water vapour also condenses on dust and other particles in the air, giving **fog**. This makes driving dangerous.

Water on roads freezes to **ice** as the Sun goes down.

Ice and frost mean animals have trouble finding food.

Pipes may burst and homes may get flooded.

Your turn

1. Write this out, using the correct word from each pair.
   Low pressure means air is rising/falling. It is a sign of fine/unsettled weather. The lower the pressure the calmer/stormier the weather will be. High pressure brings clear/cloudy skies which means very hot/cold weather in summer and very warm/cold weather in winter.

2. Explain *why* there is usually no rain, when pressure is high.

3. For some kinds of work, long spells of high pressure weather can cause problems. Give three examples – and give the season for each example.

4. Low pressure weather can cause people problems too. Give four examples.

5. It's August. Air pressure is high. You are going camping. List four items you'll pack, to cope with the weather.

6. Explain how each forms: **a** fog   **b** frost   **c** dew

7. The air pressure chart on page 74 was for a day in July.
   a. Define: **i** isobar **ii** millibar (Glossary?)
   b. State the air pressure at X. Include the unit!
   c. Describe the weather at: **i** Y **ii** Z

75

5.4 Why is our weather so changeable?

Here you'll learn why the weather in the UK can change so fast.

Our changeable weather

Our weather can be warm and sunny one day. Cool and wet the next. Why? Because the UK is in a zone where warm and cold air continually mix, to try to even out Earth's uneven heating.

Earth's two mixing zones lie 30° – 60° north and 30° – 60° south of the Equator. They correspond to the Ferrel cells, which you met on page 72. We also call them the **mid-latitudes**. Look at **A**.

And that's why our weather is so changeable. A lot of mixing goes on around us. Different **air masses** come our way.

What's an air mass?

An air mass is a huge block of air with the same temperature and moisture content across it. It might be very warm and dry, or cold and damp. It can be thousands of square kilometres in area.

Air masses form in high pressure regions where the air may sit still for days. As it sits, the air takes on properties from the surface below it.

The air mass then moves towards an area at lower pressure. The air may change on the way. For example it may pick up more moisture over water.

Where do air masses come from?

Look at **B**. The arrows show five air masses that reach the UK, and where they come from. Each affects our weather.

A

▲ Look at the UK. It's in the mid-latitudes, which correspond to the Ferrel cell. The UK stretches from 50°N to around 60°N.

▼ Air mass 4 on map B has arrived in the UK. In winter it brings us very cold weather and heavy snow – and in summer, warm or hot weather.

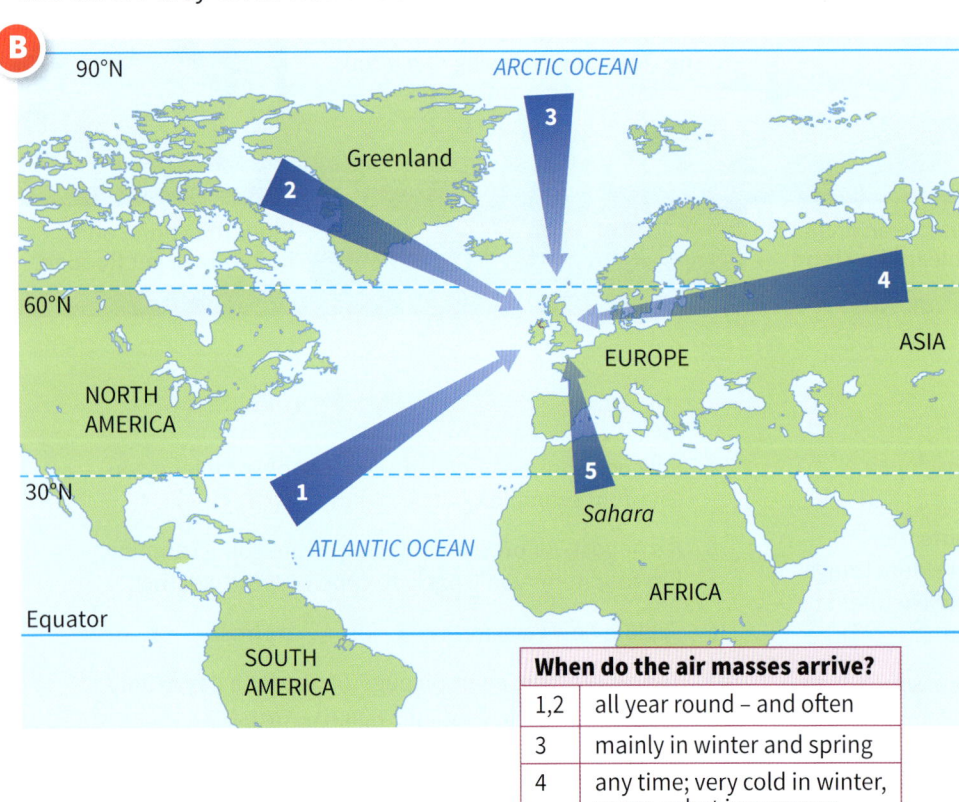

When do the air masses arrive?	
1,2	all year round – and often
3	mainly in winter and spring
4	any time; very cold in winter, warm or hot in summer
5	mainly in summer

Weather and climate

How do air masses affect our weather?

Let's look at two air masses from **B**, to see what weather they bring.

Air mass 1 in **B** forms over warm water in the Atlantic Ocean. Warm air can take up lots of moisture, so it's damp.

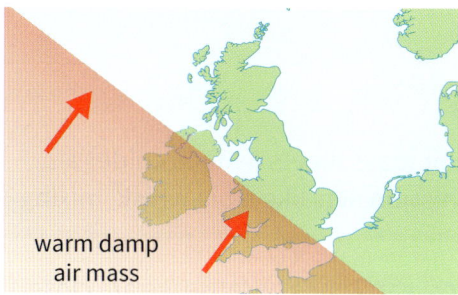

When it reaches the UK it brings us warm weather in summer, and mild weather in winter.

When it meets high land it is forced to rise. So it cools – and its water vapour condenses, giving lots of rain.

Air mass 5 in **B** forms over North Africa and the Sahara. So it's very warm and very dry.

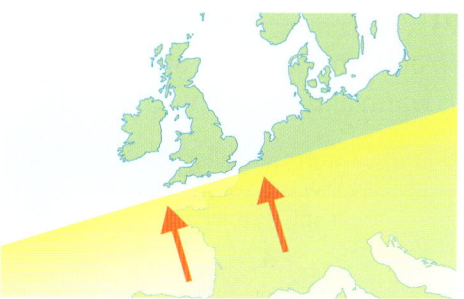

It may come our way in summer. It does not spend long over water, so it's still dry when it reaches us.

It brings us heat waves – hot days (over 30 °C) and warm nights. Don't forget your sunscreen!

Fronts

We can show an air mass on a weather map. Its leading edge is called a **front**.

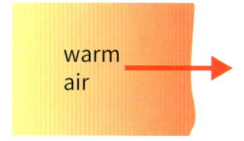

A **warm front** means a warm air mass is arriving.

It is shown like this on a weather map. (Warm red!)

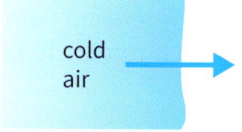

A **cold front** means a cold air mass is arriving.

It is shown like this on a weather map. (Cold blue!)

Your turn

1. The UK lies in Earth's *mid-latitudes*.
    a. State where the mid-latitudes are.
    b. Explain why our location brings us changeable weather.
2. What are *air masses*, and why do they matter to us?
3. Use the numbers in **B** to answer these questions.
    a. Which air masses are always cool or cold? Explain!
    b. Which air mass is the warmest and driest? Why?
    c. Which one is likely to be the dampest? Explain why.
    d. Look at page 69. A winter scene like this may link the UK to northern Russia. Explain why.
    e. What weather might air from Russia bring us in summer?
    f. What sort of weather will air mass 3 bring? Give reasons.
4. Air masses may change on their way to us. The paragraph below is about air mass 2 in **B**. Copy and complete it, using the words in brackets.

    When it forms in the Arctic, this air mass is _____ . It is also _____ because the temperature is too _____ for much _____ to evaporate. But as it travels towards the UK, it gains warmth from the ocean, and picks up _____ . It brings us _____ weather with heavy _____ and _____ in between.

    (sunshine   moisture   cool   water   cold   showers   dry   low)

5. a. What is: i a warm front?   ii a cold front?
   b. Draw symbols for warm and cold fronts. Beside each, write the words *warmer* and *colder* in the correct places.

77

5.5 What's a depression?

A clash of air masses can create weather systems called depressions. Find out about them here.

It's a weather system!

In Unit 5.4 you learned about **air masses**.

A **depression** forms when a warm moist air mass meets a much colder air mass. They don't mix easily. But the depression acts like a big whisk, mixing the warm and cold air together.

Around 100 depressions cross the UK every year. That's because we are close to the boundary of the Polar cell, where warm moist air and cold polar air meet. In winter, when the cold air is coldest, depressions can become big **storms**.

The life cycle of a depression

Imagine you're looking down from the top of the troposphere at the boundary between two air masses. One is warm and moist, the other cold.

▲ This satellite image for 9 August 2019 shows a depression reaching the British Isles. (Check the blue map outlines.) X: the warm front. Y: the cold front. Z: the clouds that formed as warm air rose.

A depression starts when a wedge of warm moist air pushes into the cold air mass. Warm air is lighter, so it starts rising. Air pressure falls.

➡ The warm air spirals upwards. (It spirals because of the Coriolis effect.) Cold air rushes in below, to replace it, as wind!

➡ The warm rising air cools. Its water vapour condenses. Clouds form, then rain. So a depression brings wind and rain. It may be very stormy!

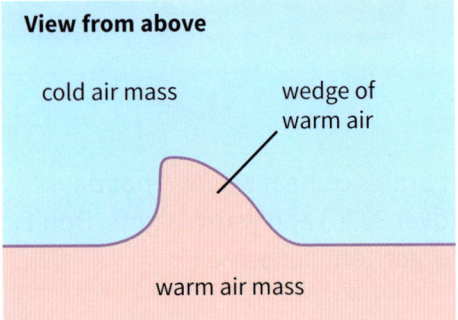

The depression travels east across the UK, bringing its weather along. It is driven by a wind high in the troposphere, the **polar jet stream**.

➡ As the depression moves, more and more of the warm air is pushed upwards. The cold front is catching up with the warm front.

➡ Now all the warm air has been lifted off the ground. It cools. The wind dies down as the depression dies away. Rain may continue for a while.

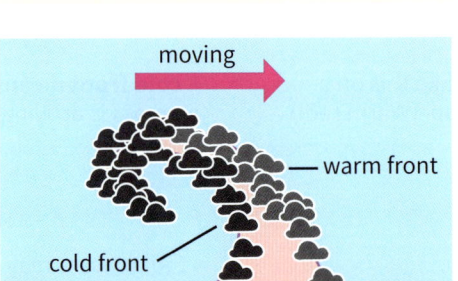
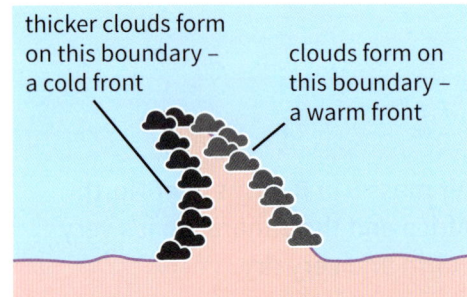

A depression can take around three days to cross the UK. It does its job well, mixing warm and cold air. But we have to suffer the weather!

Weather and climate

A stormy August

August is a holiday month. School is off. It's time to relax and enjoy the summer weather.

But not in 2019. Much of August was wet and windy. Because the polar jet stream – that powerful wind at the top of the troposphere – steered depression after depression our way.

The weekend of 9 August was the worst. Events had been planned all over the UK, and tickets sold. But a depression arrived, bringing very low pressure and very stormy weather, with gales and heavy rain.

One by one, events were cancelled. Music festivals. A kite festival. Dog shows. Horse trials. Sailing races. Surfing competitions. Balloon ascents.

Around the country, roads were flooded, trees were blown down, trains were delayed … and holidays were spoiled. Better luck next year?

▲ *Penzance, Cornwall, 9 August 2019. Gale-force winds drive waves against the sea wall.*

A depression on a weather chart

Look at **C**. It is a weather map, and it shows a depression moving across the British Isles.

The leading edge of the wedge of warm air is shown as a warm front. The cold air that follows the wedge is shown as a cold front.

Look at Z. Here the 'front' symbols alternate. This tells us that the cold front has caught up with the warm one in this area. It has pushed under the warm front and lifted it right off the ground. This is called an **occluded front**.

In time, all the warm air will have risen and cooled, and the depression will die away.

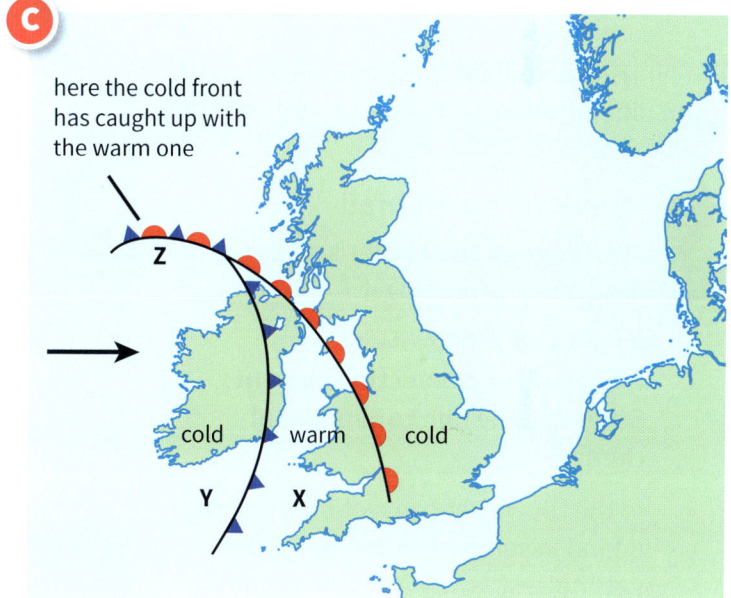

Your turn

1. Define this weather term: *depression*. (Glossary?)

2. Satellite image **A** was taken on 9 August 2019.
   a. Name the weather system it shows.
   b. Why are there so many clouds? Explain.
   c. Why do the clouds form a spiral? (Page 78 has a clue.)
   d. Depressions start over the Atlantic Ocean, and move east across the UK. What makes them move?

3. a. Satellite images help weather forecasters a lot. Why?
   b. It is 9 August 2019 – a Friday. Using **A** to help you, write a weather forecast for the UK for the weekend. Say also when you expect the weather to improve. At least 5 sentences!

4. In winter, depressions often arrive as big storms, with very low pressure.
   a. Describe the weather you'd expect in a big winter storm.
   b. Write a weather warning about this storm, for:
      i motorists   ii people living on the coast (like in **B**?)

5. a. Look at **C**. It is warmer at X than at Y. Explain why.
   b. What is happening at Z, in **C**?

6. Outline how a depression forms, and the weather it brings, for a class of 9-year-olds. Not more than 30 words.

7. Design an umbrella that will cope well in a depression. (Not like the one on page 70!) Add notes to explain your design.

5.6 More about rain ... and clouds

 Here you'll learn about three kinds of rainfall, and two kinds of cloud.

Rain: a reminder

All rain forms in the same way.

The Sun warms the water in the oceans, and rivers, and lakes.

Some water evaporates to form an invisible gas called **water vapour**. It goes into the air.

When air rises, it cools. So the water vapour **condenses** to give **clouds** of water droplets.

Droplets join to form bigger drops. When these grow heavy enough, they fall as **rain**.

But air rises for different reasons. So rainfall is given different names. Let's look at them now.

1 Convectional rainfall

This diagram shows air rising because the ground is heating it.

It rises as currents of warm air. We call these **convection currents**. So we call the rain **convectional rainfall**.

In the UK we get convectional rainfall inland in summer, where the ground gets hottest, away from the cooling effect of the sea.

2 Relief rainfall

Wind is just moving air.

When the wind meets a line of high hills or mountains, there's only one way to go – up! So the air rises and cools, and we get rain. We call it **relief rainfall**.

In the UK the prevailing wind is a moist south west wind from the Atlantic. So we get lots of relief rainfall on the high land on the west coast.

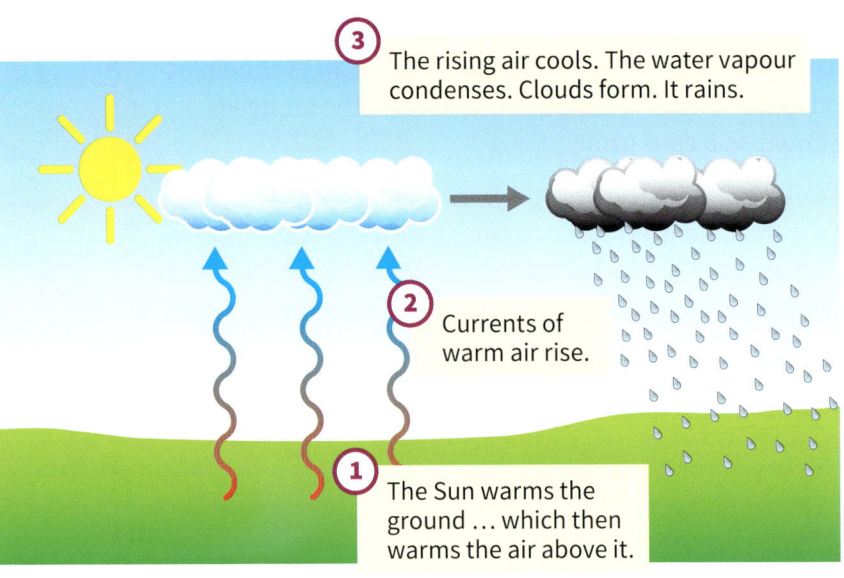
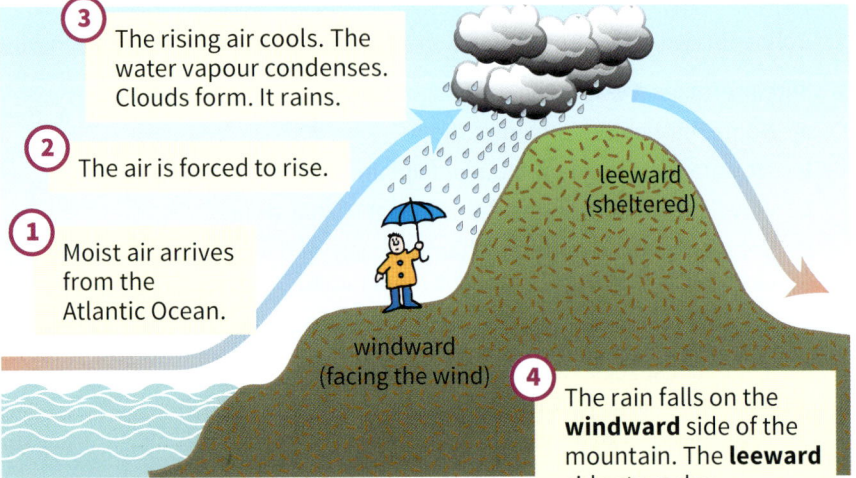

Weather and climate

3 Frontal rainfall

When a warm moist air mass meets a cold air mass, a weather system called a **depression** may develop, bringing wind and rain. (Unit 5.5.)

It rains because the warm air rises over the cold air, and its water vapour condenses.

The rainfall is called **frontal rainfall**. We get about 100 depressions a year in the UK, so it's our most common type of rainfall.

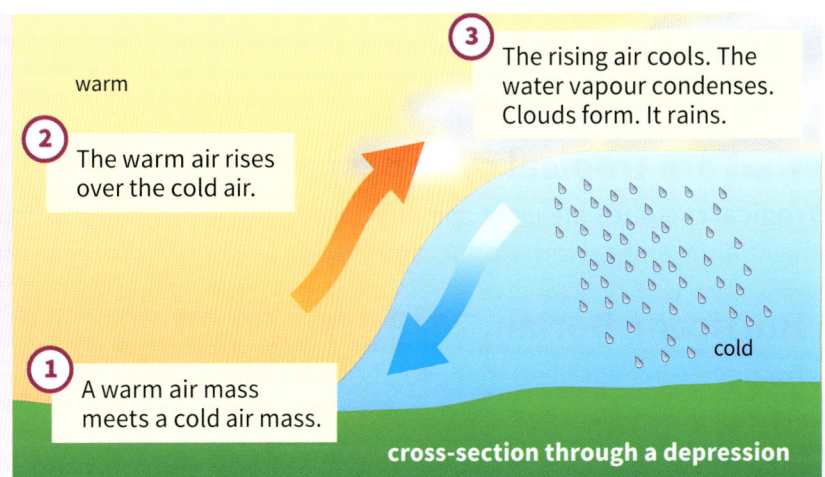

1. A warm air mass meets a cold air mass.
2. The warm air rises over the cold air.
3. The rising air cools. The water vapour condenses. Clouds form. It rains.

cross-section through a depression

Different types of cloud

There are different types of cloud too. Here are two. There are lots more!

Cumulus clouds

These fluffy clouds are a sign that warm air is rising in convection currents. For example when the ground heats the air quickly on a hot day. They can give short showers. But on a very hot day they can grow into tall dark thunderclouds, which bring very heavy rain.

Stratus clouds

These are blankets of dull cloud. They form when air rises more slowly, over a wide area – for example when a warm air mass meets a colder one. Stratus clouds bring drizzle, but not heavy showers.

Your turn

1. a Why is *relief rainfall* called that? (Glossary?)
   b Turn to page 139. Which is more likely to have relief rainfall: Princetown in south-west England, or Norwich in the east of England? Explain your answer.
2. a The most common type of rainfall in the UK is …?
   b Why is it called that? (Page 77?)
3. To form clouds, two things are *always* needed. Which are they? Choose from this list:

   wind   cooling air   mountains   hot sun
   warm ground   water vapour
4. a Can clouds form in the dark? Explain.
   b Do clouds always lead to rain? Explain.
5. The photo on the right was taken on a warm afternoon. Two hours earlier, there were no clouds in the sky.
   a Which kind of clouds are these?
   b Suggest a reason why they formed.

6 August, 3 pm

6. a The clouds in the photo above might lead to a short light shower. Name the type of rainfall it would it be.
   b It was much hotter in this place later in the week. Tall dark clouds formed, like the ones on page 75. Describe what the weather would be like then.
7. Why is rain important? Answer in any way you choose: bullet points, paragraphs, a spider map, drawings, a poem …

5.7 What's a tropical cyclone?

Here you will learn about Earth's deadliest storms: tropical cyclones.

Did you know?
- Hurricanes are named in alphabetical order through the year …
- … with alternate male and female names.

What are tropical cyclones?

Tropical cyclones are giant spinning storms. They are called **hurricanes** in the Atlantic and eastern Pacific, and **typhoons** and **cyclones** in other places.

Hurricane Dorian, 2019

For days, Gina has been trying to phone her son. She keeps trying. There is nothing else she can do.

Gina is in Florida, visiting her sister. When she arrived, Hurricane Dorian was already heading for the Bahama islands, where her home is, and where her son lives with his wife and baby.

All day she listens to the news. They say Great Abaco, her island, is wiped out. Dorian made landfall there on September 1, with winds of over 320 km an hour. A storm surge over 7 metres high swept inland. Wells filled with seawater. Rain lashed down.

And worst of all, Dorian stalled. It lingered over the Bahamas for two days, moving at just over 1 km an hour. Two days of hell.

Now there is no drinking water or power or food in Great Abaco. There is a stench of dead bodies. But Gina has not lost hope. Aid is pouring in. Survivors are being rescued. She prays she will soon hear from her son.

Meanwhile Dorian is moving up the east coast of the USA. It has weakened to a storm. Soon it will head out into the Atlantic. It will leave behind it 80 deaths – most in the Bahamas – and 4.7 billion dollars of damage.

A ▲ Somewhere in here on Great Abaco are the ruins of Gina's home.

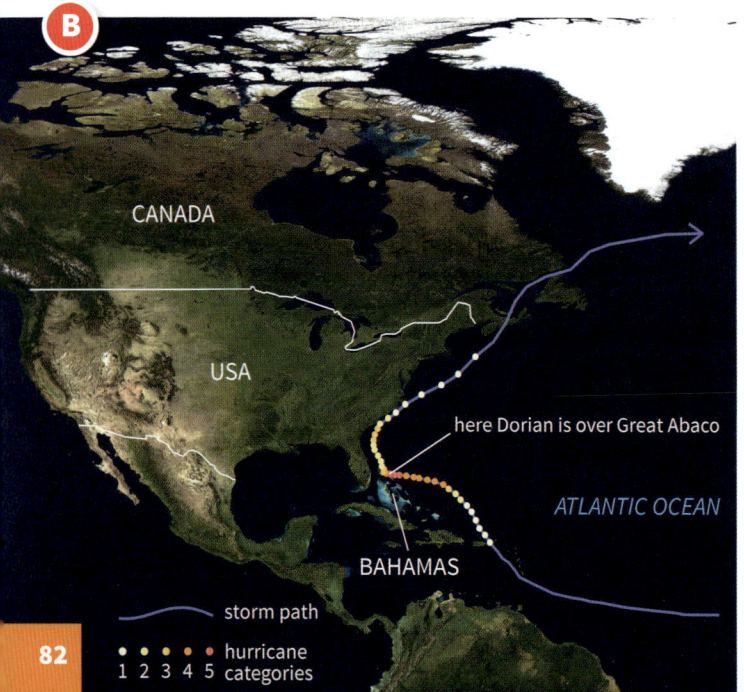
▼ Dorian grew from a storm that began in warm Atlantic waters on 24 August. When the winds reached 119 km an hour, it was classed as a Category 1 hurricane and named 'Dorian'. By landfall in the Bahamas on 1 September, it had strengthened to category 5, with winds of over 320 km an hour.

B

▼ This satellite image shows Dorian over the Bahamas, where it stalled for two days. **X** is the **eye**, where all is calm. It is surrounded by the **eyewall** (**Y**), a band of dense cloud about 16 km thick. Below the eyewall the storm is at its most violent. The bands of cloud such as **Z** are called **rainbands**.

C

82

Weather and climate

The life cycle of a tropical cyclone

Tropical cyclones are Earth's most violent storms. Their job is to carry heat away from warm seas near the Equator! Let's see how.

They start in a tropical sea, with water at least 26 °C, down to at least 50 m. And at least 5° from the Equator, to allow spin. (Coriolis!)

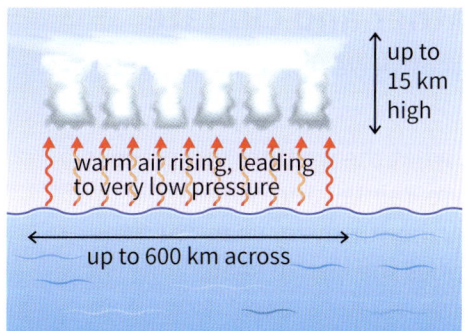

Warm moist air rises over the warm water, and tall thick clouds form. More air spirals in as wind, to replace the rising air.

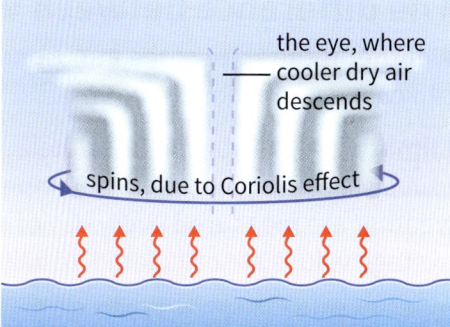

The clouds merge into bands, and the structure starts to spin. It's a hurricane! It spins faster and faster, around a calm clear centre: the eye.

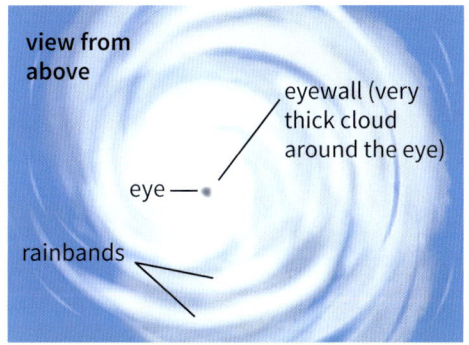

Here it is from above. Below, warm moist air continues to spiral in as wind, feeding it. Wind speeds can reach over 250 km/hour.

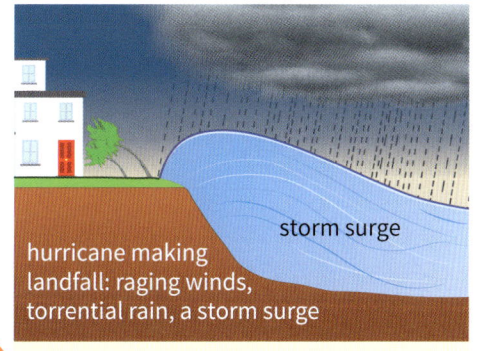

The hurricane moves as it spins, steered by external winds. It may move over land, pushing a storm surge of seawater before it.

On land, the hurricane is cut off from the warm moist air it needs. It weakens to a storm, which will eventually fade away.

Luckily, tropical cyclones (hurricanes) don't reach the UK. The colder ocean around us would starve them of the warm moist air they need. But sometimes the remains of hurricanes do cross the Atlantic to us, as much weaker storms.

Your turn

1. a What is a *tropical cyclone*?
   b Explain why the word *tropical* is used in its name.
   c State three other names that are used for it.

2. Explain why this aspect of a hurricane is dangerous:
   a the winds   b the rain   c the storm surge

3. Explain these facts about hurricanes:
   a They move, and change direction (as in **B**).
   b They weaken rapidly over land.
   c They can do damage over a very large area.
   d From above, the clouds look like a big swirl (as in **C**).

4. Imagine you took photo **A**. Describe the scene in front of you.

5. Dorian stalled over the Bahamas. Suggest a reason.

6. Describe how Dorian's strength and route changed after it moved on from Great Abaco. (See **B**.)

7. The UK does not get hurricanes. Why not?

8. The UK does get depressions, which can bring very stormy weather. Compare photo **A** on page 78 with photo **C** on page 82. What similarities do you notice? What differences?

5.8 Climate and climate factors

▲ *Dakar, the capital of Senegal in Africa.*

Here you'll learn what climate is, and why it varies so much from place to place.

The difference between weather and climate

Weather is the state of the atmosphere at any give time. As you know, it can change from hour to hour.

Climate gives the big picture. It tells you what the weather in a place is *usually* like, in a given month. Temperature, rainfall and other aspects of weather are **measured** every day of every month, every year. Then the average values are calculated, usually across the last 30 years.

The climate in different places

The climate varies from place to place – and also from month to month.

Look at table **A**. Dakar is the hottest and wettest place in August. And London is far colder in December than in August.

A

Place	Month	Average temp (°C)	Average precipitation (mm)
North Pole	August	0	15
Dakar, Senegal	August	27	182
London, UK	August	18	53
London, UK	December	6	58

The factors that affect climate

Why is the climate so different in different places, and even in different months? Let's look at the factors that affect it.

1 Latitude – the main factor

Latitude means how far a place is north or south of the Equator. It is measured in degrees.

As you saw on page 68, Earth is heated by sunlight from the Sun. But unevenly, because Earth is curved. (Check **B**.) In general, the further a place is from the Equator the cooler it is. So overall, latitude is the main factor affecting temperature, which is a key aspect of climate.

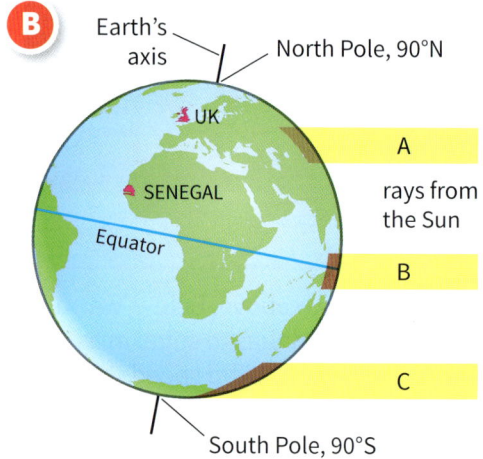

▲ *Beam C spreads its energy over a larger area than A or B. So this area is coolest.*

2 Earth's tilt plays a big part

So far, we have ignored Earth's tilt. Let's look at it now.

Earth travels non-stop around the Sun. Its axis is tilted it travels. That tilt gives us our **seasons**. It's the reason why our climate is so different in summer and winter.

Look at **C**. It shows one full orbit of Earth around the Sun, which takes a year.

In June, the top half of Earth, where we live – the **Northern Hemisphere** – is tilted *towards* the Sun. So we get warmer. We have summer.

But by December, it is tilted *away* from the Sun. So we get colder. We have winter!

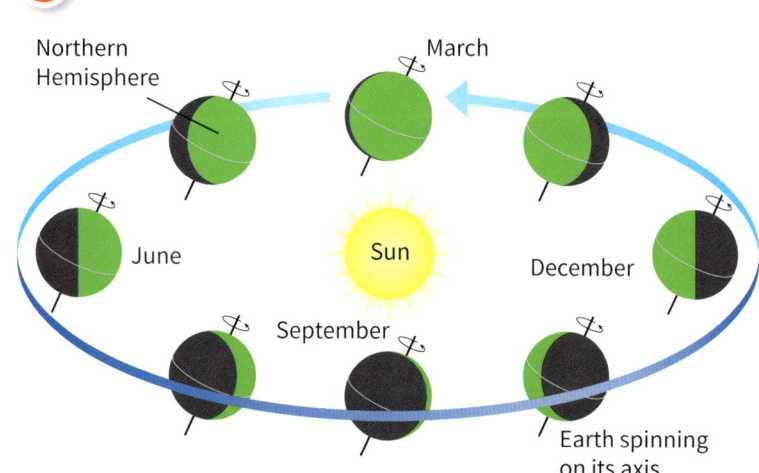

Weather and climate

3 The movement of heat around Earth plays a big part

In Unit 5.2 you saw how heat gets moved around Earth, to even it out. This leads to prevailing winds, and high and low pressure belts, and ocean currents. They all affect the climate in a place. Look:

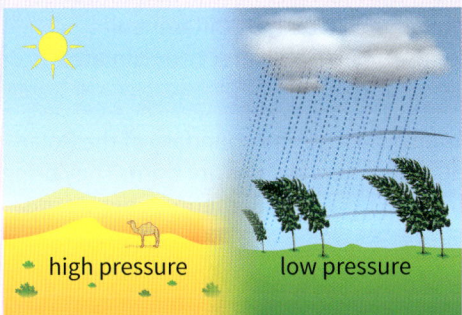

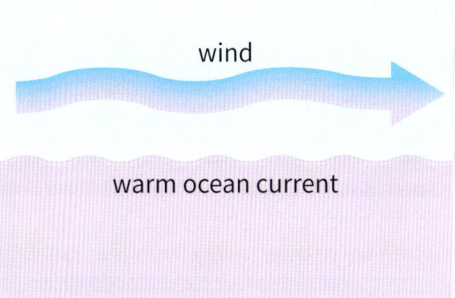

Earth's high and low pressure belts.
In a high pressure belt the skies are usually clear, and there's no rain. Low pressure means wind and rain.

Prevailing winds.
For example if the prevailing wind blows in from the ocean, it will bring water vapour – which means rain!

Ocean currents.
A warm current warms the wind, which then warms the places it blows to. A cold current does the opposite.

4 And the last two factors …

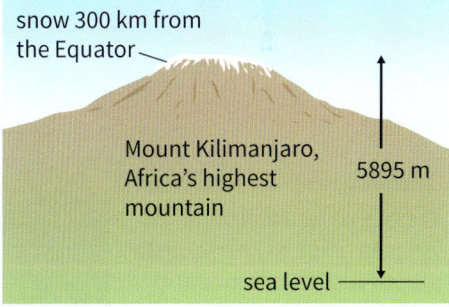

Distance from the sea.
The sea is slower to warm up and cool down than land. So it keeps the coast cool in summer – and warm in winter!

But on a big land mass, far from the sea, the land heats up fast and gets very hot in summer. Then cools fast and gets very cold in winter.

Height above sea level (altitude).
The higher you are above sea level, the cooler it is. The temperature falls by about 1 °C for every 100 metres.

So the climate in a place is a balance between all these factors. It's complex! And sometimes another factor may have more influence than latitude.

Your turn

1. Define *climate*. (Glossary?)
2. Is this statement about weather, or about climate?
   a. March is usually the wettest month in London.
   b. It was very foggy on the motorway last night.
   c. The temperature in Singapore is rarely below 23 °C.
3. Write sentences to show that you understand these terms:
   a. latitude   b. altitude   c. prevailing winds
4. Look at table **A**. Explain why:
   a. Dakar is warmer than London in August
   b. London is colder in December than in August
   c. it's only 0 °C on average at the North Pole in August
5. Heat gets moved around Earth, as you saw in Unit 5.2. With the help of set **3** above, explain how this affects climate.
6. Here are the answers. What are the questions?
   a. *Because the sea heats up more slowly than land does.*
   b. *Because the sea cools down more slowly than land does.*
   c. *Because the air gets thinner, so can hold less heat.*
7. Other factors can win out over latitude, in influencing the *temperature* in a place. Find one example on this page.
8. Identify three factors that bring some places more *rainfall* than others. (Does altitude play any part?)

5.9 So what's the UK's climate like?

Here we look more closely at the UK's climate, and how the different climate factors affect it.

The UK and those climate factors

In Unit 5.8 you met the climate factors. Now let's see how they apply to the UK.

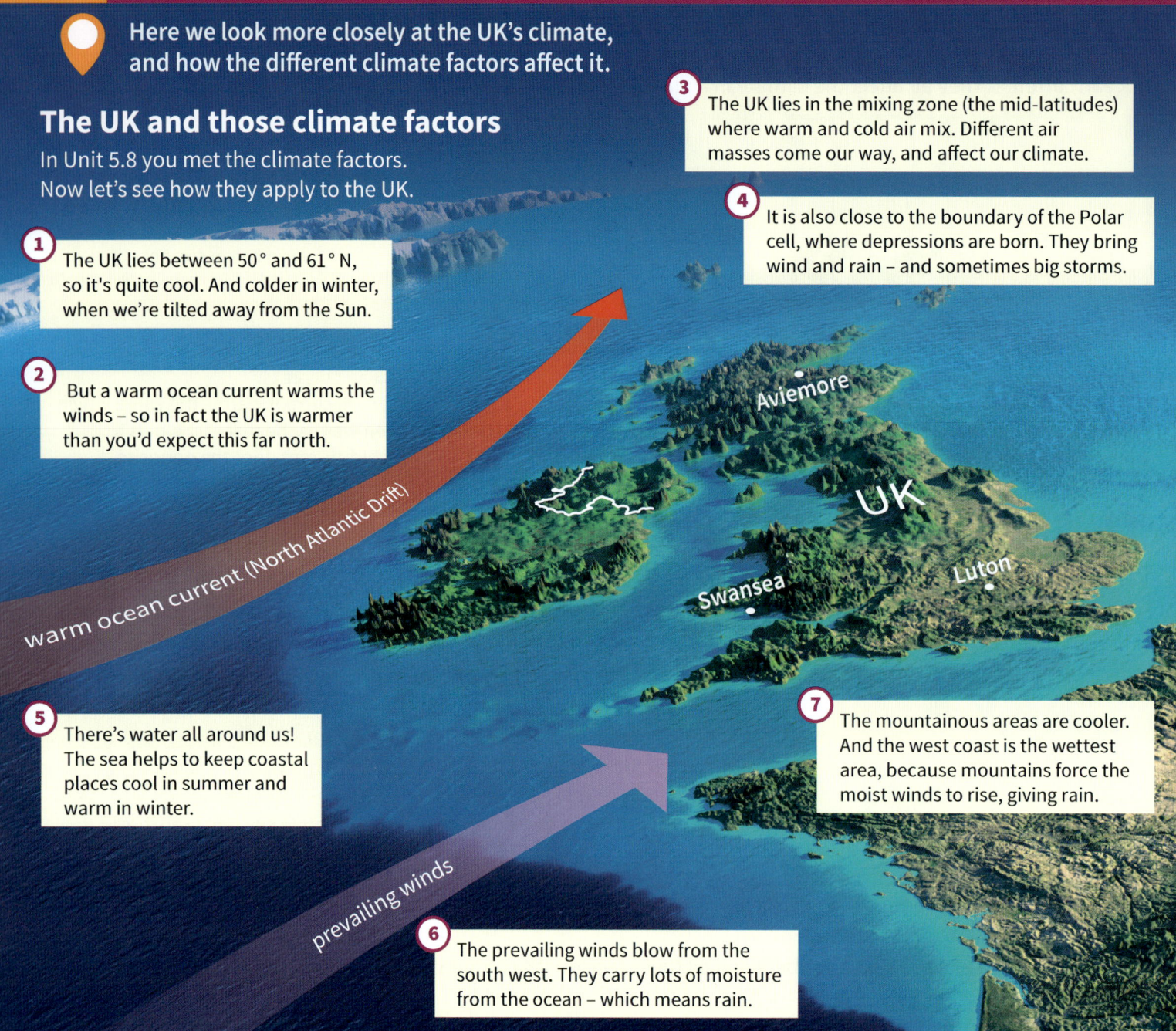

1. The UK lies between 50° and 61° N, so it's quite cool. And colder in winter, when we're tilted away from the Sun.

2. But a warm ocean current warms the winds – so in fact the UK is warmer than you'd expect this far north.

3. The UK lies in the mixing zone (the mid-latitudes) where warm and cold air mix. Different air masses come our way, and affect our climate.

4. It is also close to the boundary of the Polar cell, where depressions are born. They bring wind and rain – and sometimes big storms.

5. There's water all around us! The sea helps to keep coastal places cool in summer and warm in winter.

6. The prevailing winds blow from the south west. They carry lots of moisture from the ocean – which means rain.

7. The mountainous areas are cooler. And the west coast is the wettest area, because mountains force the moist winds to rise, giving rain.

Comparing three places in the UK

The UK is not a big country. But even within it, the climate varies. Table **A** has data for the three places marked on the image above. You'll use it in *Your turn*.

A Climate data for three places in the UK

		Jan	Feb	Mar	Apr	May	Jun	Jul	Aug	Sep	Oct	Nov	Dec
Average temperature (°C)	Aviemore	−0.7	0.3	3.9	4.4	7.5	13.2	16.1	15.2	10.0	6.9	4.0	0.7
	Swansea	4.4	4.7	6.6	9.3	12.2	15.6	16.8	16.5	14.3	11.1	7.8	5.7
	Luton	3.0	3.3	6.0	8.6	12.1	15.1	17.4	16.9	14.2	10.5	6.4	3.9
Average precipitation – rain or snow (mm)	Aviemore	90	60	69	47	59	60	64	79	79	87	85	86
	Swansea	125	83	90	66	68	70	70	91	101	118	130	132
	Luton	59	41	51	51	52	55	49	55	57	59	59	63

A climate graph

We can show climate data in a table, as in **A**.
We can also show it on a **climate graph**.

B is a climate graph for Swansea. The red line shows average temperature. The blue bars show rainfall.

For June, the average temperature is 15.6 °C.
(Read it at X, the mid-point for June.)

And June has 70 mm of rainfall.
(So if it did not drain away, this rain would give a layer of water 7 cm deep in June, all over Swansea!)

But remember, climate data are *averages*.

A climate graph for *your* place would give you an idea of what to expect this month. But to find out what the weather will be like tomorrow, you'll need to look up the weather forecast.

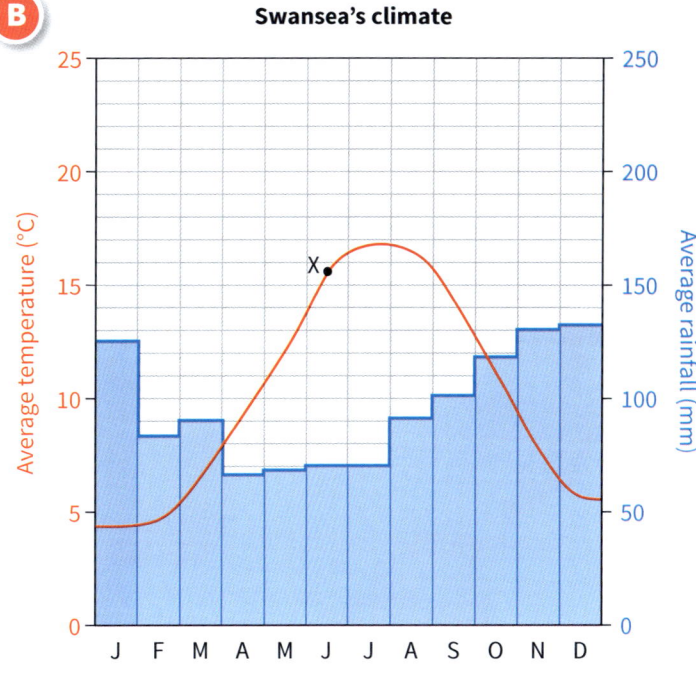

▲ *A climate graph for Swansea.*

Your turn

The map on page 139 may help with some of these questions.

1. a Between what latitudes does the UK lie?
   b Use your answer in **a** to explain why the UK is not a hot country.
   c In fact, the UK is warmer than you'd expect for its latitude, because … . Copy and complete the sentence.

2. *The ocean has an enormous impact on our climate.*
   Write a paragraph to justify the statement in italics.
   (Remember, seas are just areas of the ocean!)

3. Look at table **A**.
   a Which is the hottest month of the year in:
      i Swansea?   ii Luton?   iii Aviemore?
   b Which is the coldest month of the year in:
      i Swansea?   ii Luton?   iii Aviemore?
   c One of the three places in **A** offers outdoor skiing in January. Which one? Justify your choice.
   d *The UK has different seasons.* Do the temperatures in **A** support this statement? Explain your answer.

4. Luton and Swansea are at roughly the same latitude. Suggest a reason why:
   a Luton is usually colder than Swansea in January
   b Luton is usually warmer than Swansea in July

5. All through the year, Aviemore is colder than both Luton and Swansea. Give *two* reasons to explain this. (Page 139?)

6. Now look at the rainfall data in **A**.
   a What unit is used to measure rainfall? Give its full name.
   b i Which place in **A** gets the most rainfall every month?
     ii Give *two* reasons to explain why this place gets a lot of rainfall. (Page 139 may help.)
   c Luton gets much less rainfall than Swansea. Explain this.

7. a When do the three places in **A** usually get more rainfall?
      i in summer (June – August)
      ii in winter (December – February)
   b The UK gets more depressions in winter. Explain how this helps to account for what you found in **a**.

8. **B** shows a climate graph for Swansea.
   a What does the red line show? (Check the red axis.)
   b Which two months are:
      i hottest?   ii coldest?
   c What do the blue bars show? (Check the blue axis.)
   d Which two months are the driest?

9. **A** and **B** show the same climate data for Swansea. Which do you think is the better way to show how the climate in a place changes through the year?
   a a table of data   b a climate graph
   Justify your choice.

10. You plan to visit Swansea next week. Explain why a weather forecast will be more useful to you than either **A** or **B**.

5.10 Climates around the world

Here you will see how climate varies around the world – and in *Your turn* you'll be a climate detective.

A world climate map

A

Did you know?
- The lowest temperature ever recorded was −89.2 °C, in Antarctica.
- The highest was 56.7 °C, at Death Valley in California, USA.

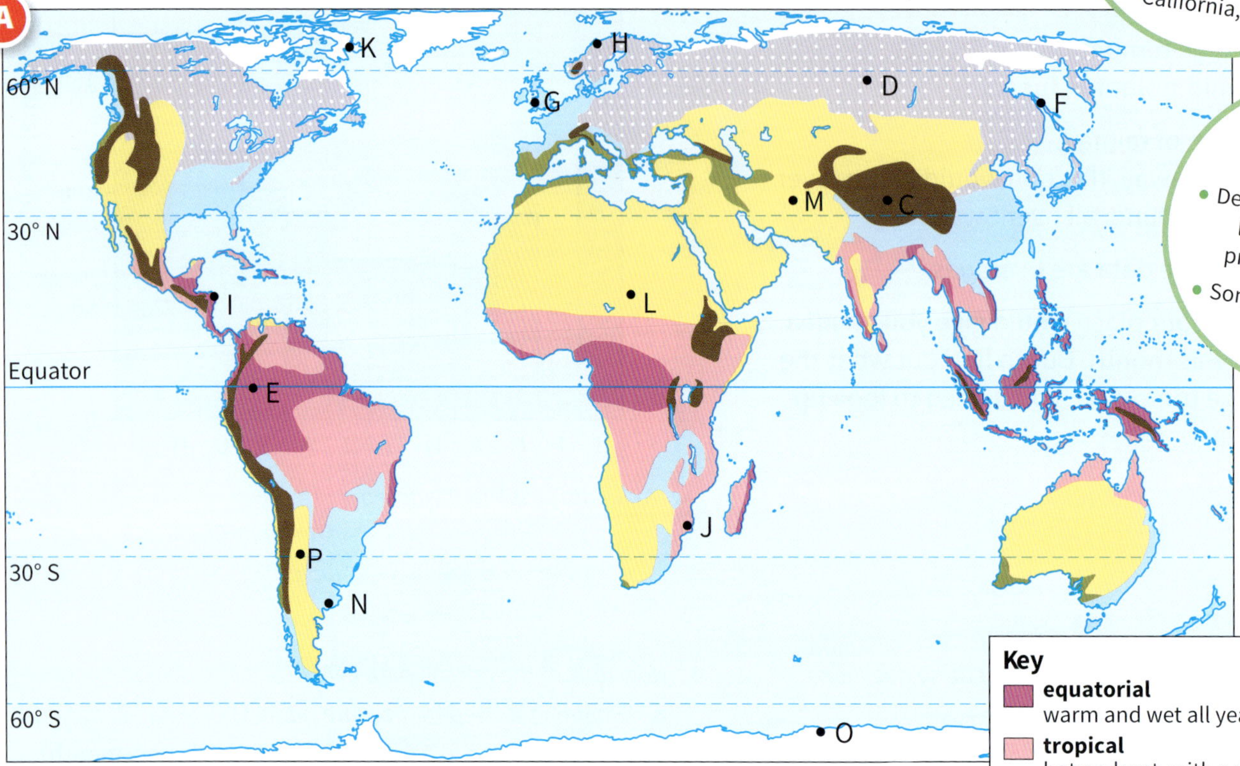

Did you know?
- Deserts are places with less than 25 cm of precipitation a year.
- Some deserts are hot, some cold.

Key
- **equatorial** — warm and wet all year
- **tropical** — hot and wet, with a dry season
- **desert** — very dry – under 25 cm of precipitation a year
- **mediterranean** — hot dry summers, warm wet winters
- **maritime** — warm summers and cool winters, wet
- **continental** — hot summers and cold winters, dry
- **polar** — very cold all year (especially in winter), and dry
- **mountain** — cold because it is high, with heavy rain or snow

Map **A** shows Earth's **climate regions**.

Let's look at the big pattern …

- In general, the further from the Equator you go, the colder the climate.
- In the equatorial regions (deep pink), it is warm and wet all year. Warm because the Sun's energy is more concentrated around the Equator. And wet because the warm air rises fast, its water vapour condenses to form thick clouds – and then the rain pours down. Every day.
- But the equatorial regions are not Earth's *hottest* regions! That's because the thick clouds stop some sunlight getting through. The hot deserts are hotter.
- Look at the desert areas. Most are centred on the high pressure belts around 30° N and S, where air sinks. (See **C** on page 73.) Sinking air means no clouds to block sunlight, and no rain. So these are Earth's driest regions, and the hot deserts are the hottest regions. (There are cold deserts too.)
- It's very dry at the poles, because it's too cold for evaporation – and these are high pressure areas.
- Look at the UK. It has a **maritime** climate overall. (From *mare*, the Latin word for *sea*.) This term reflects the big impact of the ocean on us.
- But as you saw in Unit 5.9, the climate varies within the UK – and it's the same for other countries. Mountains, ocean currents, prevailing winds, distance from a coast – all play a part in local climates.

▼ Antarctica: in a high pressure belt. They're launching a weather balloon.

Weather and climate

Your turn

1. Which two climate types in **A** are the most different? (Key?)
2. Explain where the names came from, for these climate regions:
   - a equatorial
   - b tropical
   - c polar
   - d mediterranean
   - e maritime
3. List the types of climate in Australia. (Page 141?)
4. What's the climate like at these places on map **A**?
   - a L
   - b D
   - c E
   - d J
5. Look again at **A**. Explain each of these facts. Unit 5.8 may help.
   - a It is a lot cooler at D than at L.
   - b There is very little precipitation at O. (Glossary?)
   - c Any precipitation at O is in the form of snow.
   - d It's hotter at L than at E, even though E is on the Equator.
   - e G (in the UK) and N (in Argentina) both have a maritime climate – but G is much warmer than N in July.
6. Map **B** shows prevailing winds, ocean currents, relief, and vegetation cover.
   - a Define: i vegetation ii relief
   - b How can you tell which areas are desert, on **B**?
7. Latitude is a major factor affecting climate – but not the only factor. This is about places on **A** that are at the same latitude. Give reasons to explain the facts below. (There may be more than one reason.) Map **B** may help.
   - a It's much warmer at M than at C.
   - b It's much warmer at G than at F.
   - c It's much colder at K than at H.
   - d It is much wetter at I than at L.
8. A warm ocean current warms the wind, which then warms the land it blows over. A warm wind can also take up more moisture than a cold wind.
   - a Predict the effect of a *cold* ocean current on:
     - i the temperature of the wind that blows over it
     - ii the amount of moisture in this wind.
   - b Find P on map **A**. Using **B** to help you, explain why the area around P is a cold desert.
9. This is a climate graph for a place called Frobisher Bay.

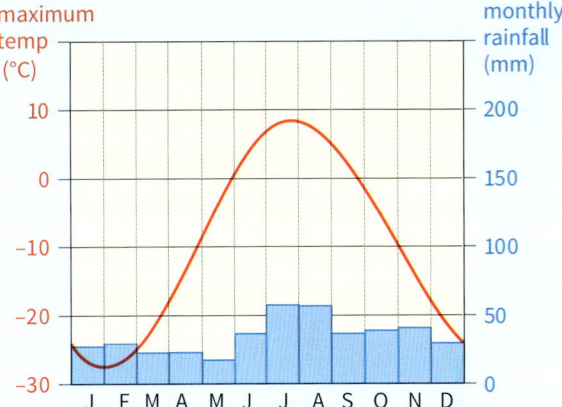

   - a Which two months are the warmest, in Frobisher Bay?
   - b Identify the driest month.
   - c In which months can the residents expect snow?
   - d Frobisher Bay is marked on **A** by a letter. Which letter is it: G, J, K, or L?

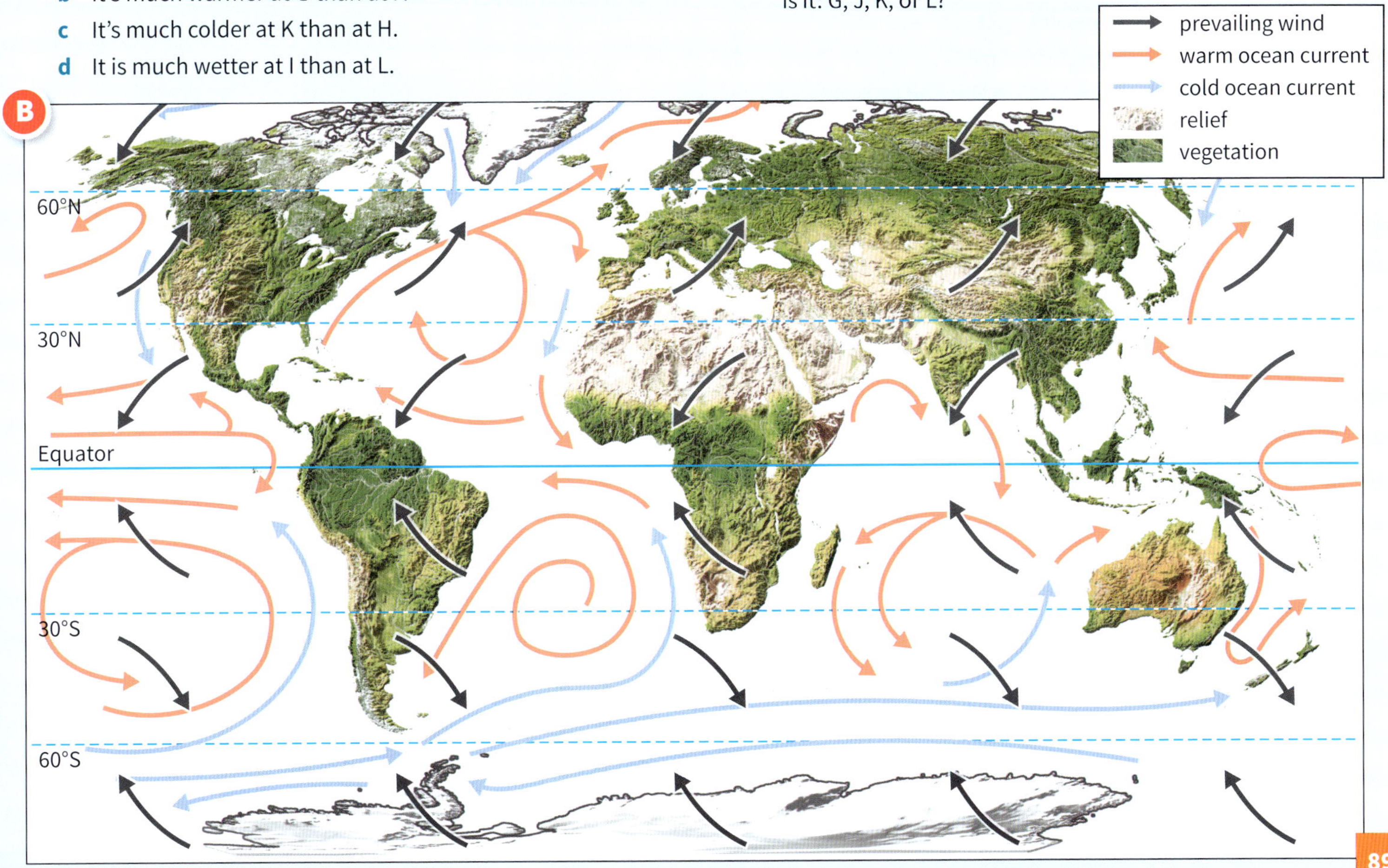

Key:
→ prevailing wind
→ warm ocean current
→ cold ocean current
   relief
   vegetation

B

5 Weather and climate

How much have you learned about weather and climate? Let's see.

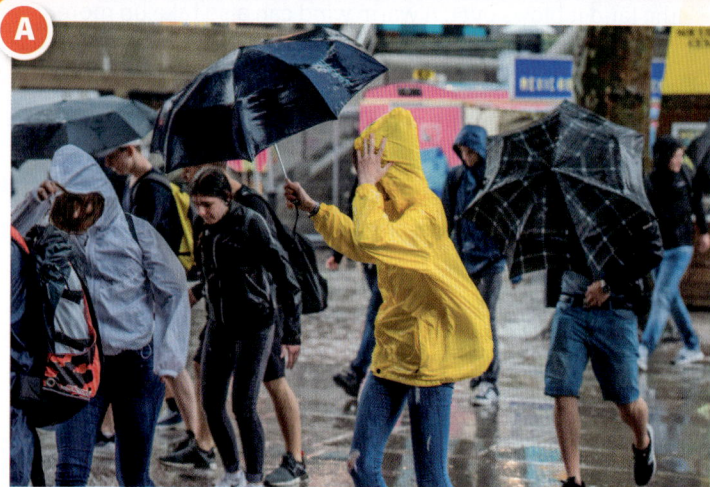

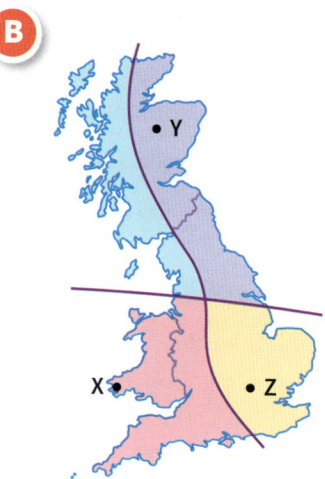

1 **A** shows a wet and windy day in the UK.
   a Name the substance that falls as rain.
   b Explain how rain forms. (You could draw a diagram.)
   c The UK gets three types of rain. State their names.
   d This is about wind. Write it out with all the words unjumbled!
      Wind is *vimong ria*. It *swolf* from an area with *gihh purreess* to one with *owl supreser*. It tries to *veen otu* the *eruspres*.
   e Name a weather system which often crosses the UK, bringing weather like that shown in **A**.

2 Copy and complete, using words from the brackets below.
   a When air rises, air pressure _____ .
   b When air sinks, air pressure _____ .
   c Low pressure leads to _____ and _____.
   d High pressure leads to _____ skies and no rain.
   e In summer, high pressure leads to _____ _____ weather. Nights are _____ because there are no clouds to keep heat in.

   (*sunny   cool   rain   rises   warm   wind   cloudless   falls*)

3 The UK lies in Earth's mid-latitudes. Our weather is changeable because different air masses come our way.
   a Where are the *mid-latitudes*?
      i Define *air mass*.
      ii One air mass arrives in the UK mainly in summer. State where it comes from, and describe how it affects our weather.
      iii Outline what happens when a warm moist air mass meets a polar air mass.
   b The mid-latitudes correspond to one of the three pairs of cells that make up the global atmospheric circulation.
      i State the name of this pair of cells.
      ii Name the other two pairs of cells.
      iii Explain why the global atmospheric circulation is necessary.

4 **B** shows Britain divided roughly into four climate zones.
   a Which place is likely to be warmer all year, X or Y? Why?
   b X and Z are at the same latitude. Explain why Z is:
      i colder than X in winter   ii drier than X
   c Make a larger copy of **B**. A quick rough sketch is fine!
   d Add these labels in the correct zones on your map. (Think about our prevailing wind direction!)
      warm summers, mild winters, wet
      cool summers, mild winters, wet
      warm summers, cold winters, quite dry
      cool summers, cold winters, not so wet
   e Write a paragraph to show that you understand the difference between *weather* and *climate*.

5 a Make a wide table with these headings:

Depression	Tropical cyclone

   b Then write the sentences below in the correct columns. Do them in order. Some belong to both columns.
   • It's called a hurricane when it occurs in the Atlantic.
   • It starts in warm tropical waters.
   • It starts at a boundary between warm and cold air.
   • The first step is air rising, giving low pressure below it.
   • The rising air leads to clouds, and then rain.
   • Wind spirals in towards the low pressure area.
   • The wind can be over 250 km/hour.
   • Bands of cloud begin to spin like a Catherine wheel.
   • It moves west to east, steered by a jet stream.
   • If it moves onto land, it will soon die away.
   • It is Earth's most intense storm.
   • Around 100 cross the UK each year.

6 *Climate plays a big part in people's lives.*
   To what extent is this true? Write at least half a page. You could think about what we wear and eat, the buildings we live in, and our work and leisure activities.

6 Climate change

6.1 Earth's climate – always changing!

Earth was formed about 4.5 billion years ago. Since then, its climate has changed continually. Find out more here.

Earth and its climate

Earth is about 4.5 billion years old. Over that time, its climate has changed continually. It has mostly been warmer than now, with no ice anywhere. But there have been many **ice ages**, when Earth cooled and ice sheets and glaciers spread.

5 million years of climate change

Let's think about Earth's climate for just a small part of its history: the last 5 million years. We'll focus on temperature. Look at graph **A**.

- The dashed line represents Earth's average temperature today. It is labelled 0 because the graph is about *changes* in temperature compared with today.

- The zig-zag line shows how the average temperature rose and fell compared with today's.

- Look at X. In this period Earth was almost always warmer than today – sometimes over 2 °C warmer.

- Look at Y. This period, from 2.6 million years ago until today, is called the **Quaternary period**. It has been almost always colder than today. Ice sheets spread from the poles each time Earth grew colder, giving **ice ages**. They retreated again as Earth warmed.

- The red part shows the last 110 000 years. The most recent ice age began around 110 000 years ago. About a third of Earth's land surface got covered in ice sheets, including much of Britain.

- Then about 12 000 years ago, Earth began to warm up. The only ice that remains today is around the poles, and up high mountains.

▲ Woolly mammoths first appeared about 400 000 years ago. They disappeared 4000 years ago – either because we hunted them, or Earth grew too warm for them, or both.

▲ In the last ice age, sea levels fell by 120 m. So sea floor was exposed as 'land bridges', allowing humans to migrate. Land bridges joined Britain to Ireland and France.

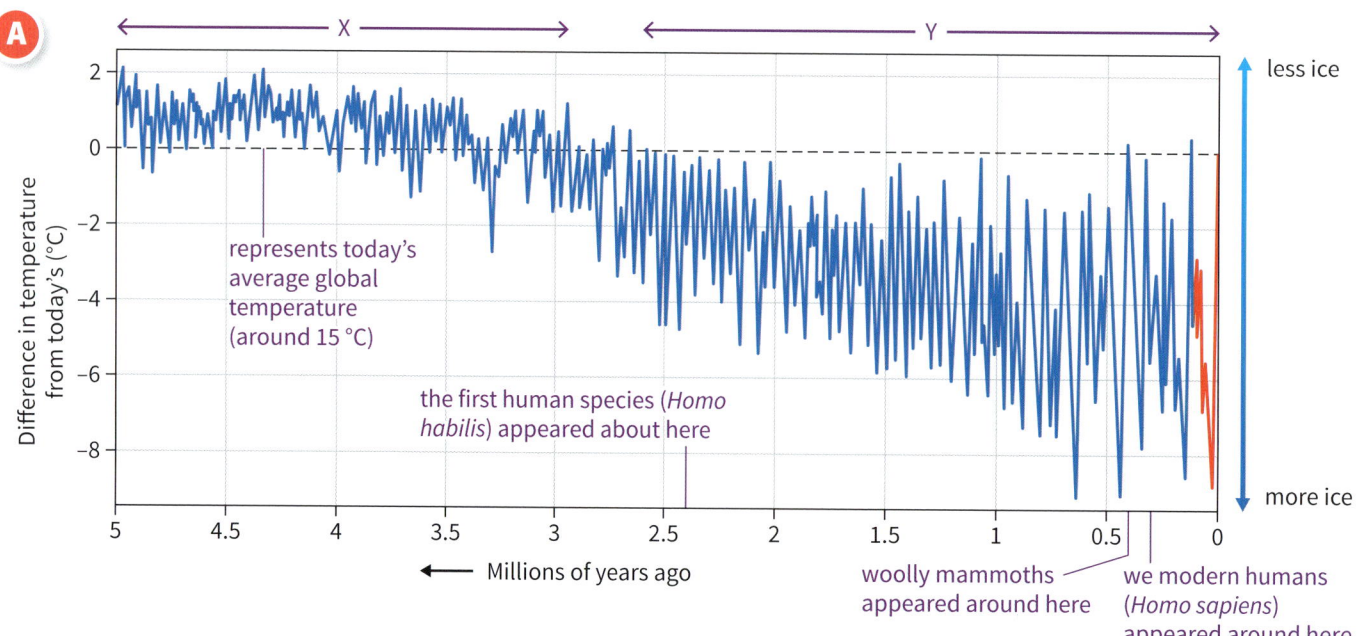

Climate change

Temperatures since the last ice age

The last ice age began around 110 000 years ago. Then, about 12 000 years ago, Earth began to warm up. The ice age ended.

But the temperature did not rise steadily. Look at the wiggly line in **B**. It shows the average temperature for the Northern Hemisphere for the last 11 000 years. The numbered notes match points 1 – 5 on the graph.

Today the average temperature in the Northern Hemisphere is around 15 °C. And it is still rising. Find out more in Unit 6.3.

Today, the temperature is still rising. How will it affect us – and other living things? You'll find out in later units.

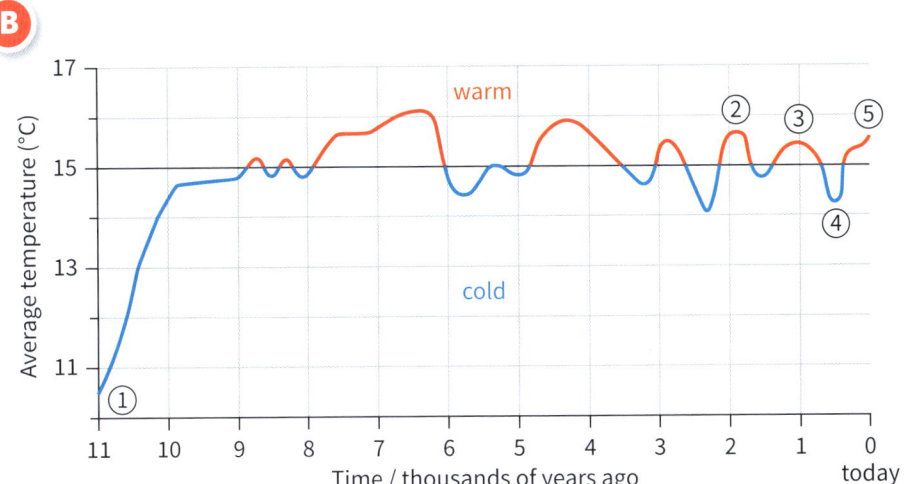

Brrr! A cold dip, around 1300 to 1850. It's called the **Little Ice Age**. In some winters the River Thames freezes over.

We modern humans (*Homo sapiens*) reached Britain around now. (France had no ice sheets in the last ice age.)

The red on the graph shows warmer periods. This one occurs in the time of the Roman Empire.

This is the **Medieval Warm Period**. The warmth allows Vikings to settle in parts of Greenland.

Your turn

1. Define these terms. (Glossary?)
   a. ice age b. Quaternary period
2. Using **A**, describe how Earth's average temperature has changed over the last 5 million years. Write at least 8 lines.
3. 5 million years ago, Earth was warmer than today. Could we humans have caused this warming? Explain. (Check **A**!)
4. Could Earth have more ice ages? Explain your thinking.
5. In the last ice age, animals were able to *walk* from icy Britain to warmer France. Explain why this journey was possible.
6. Was the Little Ice Age *really* an ice age? Explain. (y axis on **B**?)
7. Was Earth ever warmer than now, in the last 11 000 years? Give evidence from **B**.
8. List five challenges we humans would face if we were plunged into another ice age.

6.2 The climate detectives

How do we know that Earth's climate changed in the past? And what caused the changes? Find out here.

Detectives at work

We have been measuring climate properly for less than 200 years. So how can we tell about past climate changes – even millions of years ago? Scientists look for clues, like detectives do.

Looking for clues

Scientists look for clues about past climates in many places. For example:

Did you know?
- Every year, billions of tonnes of dead organisms, and particles carried by the wind and rivers, settle on the ocean floor.
- They drift down through the ocean non-stop, hour after hour after hour.

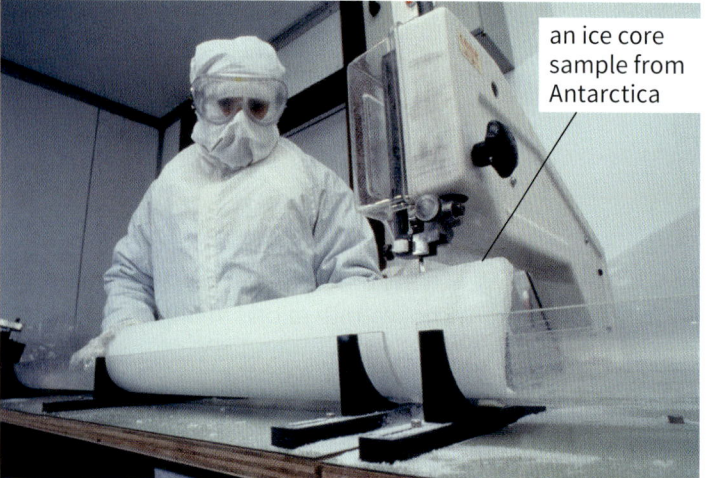

assembling the long metal tube to be drilled into the ocean floor, below the ship

Ocean sediment. It builds up layer by layer over millions of years. So it's like a history book of climate change! A core of sediment is drilled from the ocean floor …

… and studied layer by layer, using a method called **radiometric dating**. This tells scientists both the age of a layer *and* what the climate was like then.

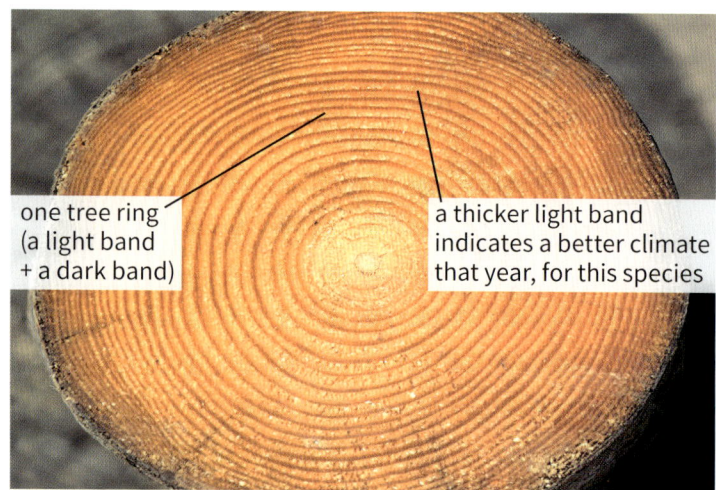

an ice core sample from Antarctica

one tree ring (a light band + a dark band)

a thicker light band indicates a better climate that year, for this species

Ice sheets. These build up in layers too, from snow. Analysis of ice cores tells scientists when the snow fell, the temperature then – and what gases were in the air.

Tree rings. Counting the rings tells us a tree's age. Their widths tell us about changing climates. You don't have to cut the tree down: you can drill a plug from a living tree.

But working out *how* and *when* climate changed are only part of the task. Scientists must also find out *why*!

The data from above, matched to data from astronomers and others, has helped scientists to find the answers. Read on …

Climate change

Factors that influence climate change

1 Changes in Earth's movements in space

Planet Earth is always on the move, taking us along for the ride. It travels non-stop around the Sun, while spinning on its axis. And its movements change over time! Look at the diagram.

- Earth's path or **orbit** around the Sun changes. So the amount of sunlight Earth receives also changes. One cycle of change takes 100 000 years.
- The tilt of its axis changes too! In cycles of 41 000 years. The bigger the tilt, the more sunlight the poles get in summer, and the less in winter. Brrrr.
- Earth's axis also swings around, in cycles of 23 000 years. This affects the amount of sunlight Earth receives.

These cycles are called the **Milankovitch cycles**. They played a major part in past climate changes – and especially the orbital cycle.

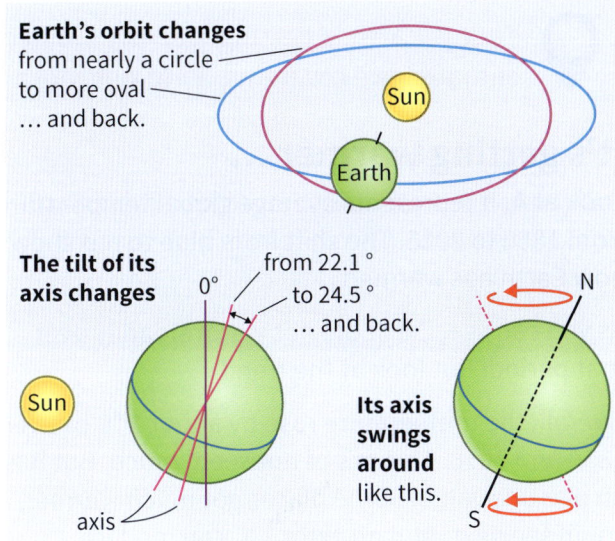

▲ Changes in Earth's movements.

2 Changes in the Sun's output

The amount of energy the Sun gives out changes too – because of magnetic activity at its surface. This activity makes the Sun shine more brightly.

The activity centres on **sunspots** that move across the Sun. Their number rises over an 11-year cycle, then falls and a new cycle starts. The cycles may affect Earth's climate – slightly. Scientists are still studying them.

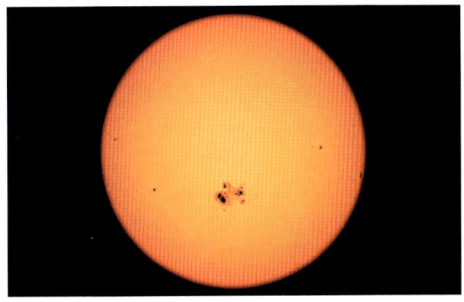

▲ Sunspots – the dark spots moving across the Sun: a sign of intense magnetic activity.

3 Volcanic eruptions

Big volcanic eruptions can blast millions of tonnes of ash and gases into the atmosphere. The ash spreads and acts like a veil, blocking out sunlight. It can cause a temperature fall for a short time, until it's washed to Earth.

But sulphur dioxide, a volcanic gas, has a much bigger cooling effect – which can last from months to several years. It forms a mist of sulphuric acid high in the atmosphere. This scatters and absorbs sunlight, keeping it from Earth.

4 Us!

Today Earth is warming – fast! And scientists say that this time *we* are the main reason. We are adding **greenhouse gases** to the air, which warm it. Some will remain in the air for hundreds or even thousands of years.

▲ Mount Pinatubo in the Philippines erupting in June 1991. Global temperatures fell by 0.5 °C for two years.

Your turn

1. **a** Describe what is going on in the first photo on page 94.
   b **i** During drilling, the long tube will fill with material. It's important not to mix this material up. Why?
   ii What can scientists learn from this material?
2. The scientist in the third photo is all covered up. Why?
3. Which do you think provides data from longer ago: tree rings, or ocean sediment? Give your reasons.
4. **a** Describe three ways in which the movements of Earth in space change over long periods of time.
   b Explain *why* these changes affect climate.
5. The factors 1 – 4 above all play a part in climate change.
   a **i** Which one affects climate for the *shortest* time?
   ii Explain how this factor affects climate.
   b Which factors are not under our control? Explain.

6.3 How is Earth's climate changing today?

Today, Earth is getting warmer. So in countries all across Earth, climates are changing. Find out more here.

It's getting warmer ...

Look at **A**. It represents average global temperatures from 1850 to 2018. The shift from blue to red shows how Earth has warmed.

As you can see, Earth did not warm steadily over that period. But look at the trend.

Overall, the temperature rose by about **1 °C** between 1850 and 2018. This might not seem much. But it is an *average* value. Some places got much warmer, as you'll see next. And an extra 1 °C can melt ice.

The warming is uneven

Look at **B**. It shows how the temperatures around Earth had changed by 2018, compared with the average for 1951 – 1980.

This warming of Earth is called **global warming**. And as the map shows, it is uneven.

Some places had warmed by less than 0.5 °C. But the Arctic had warmed by more than 2.5 °C.

What patterns can you see?

Climate change

A and **B** are about *the rise in temperature*.

But all aspects of Earth's climate are connected. So as temperature rises, the patterns of rainfall and wind and ocean currents change too. It is **climate change** in our lifetime. And it will continue.

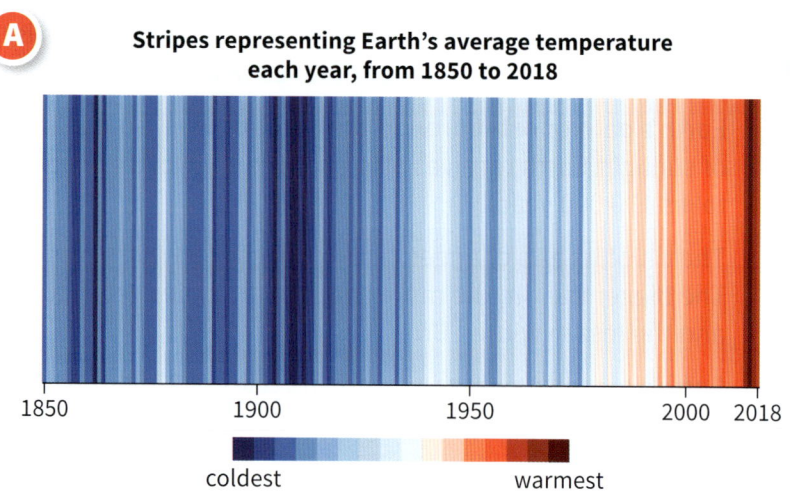

A Stripes representing Earth's average temperature each year, from 1850 to 2018

coldest — warmest

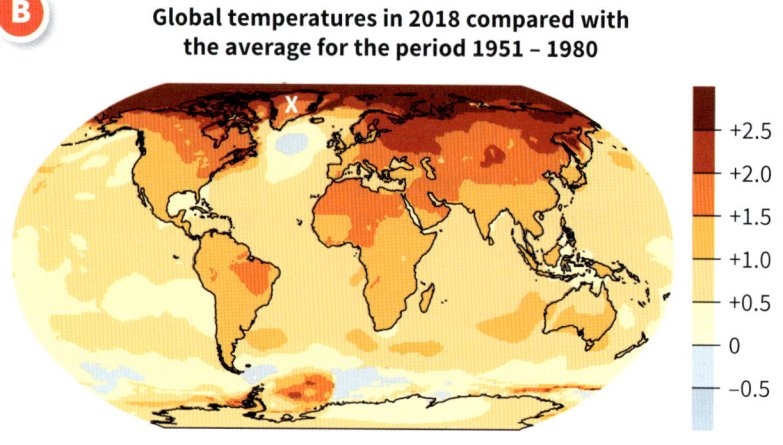

B Global temperatures in 2018 compared with the average for the period 1951 – 1980

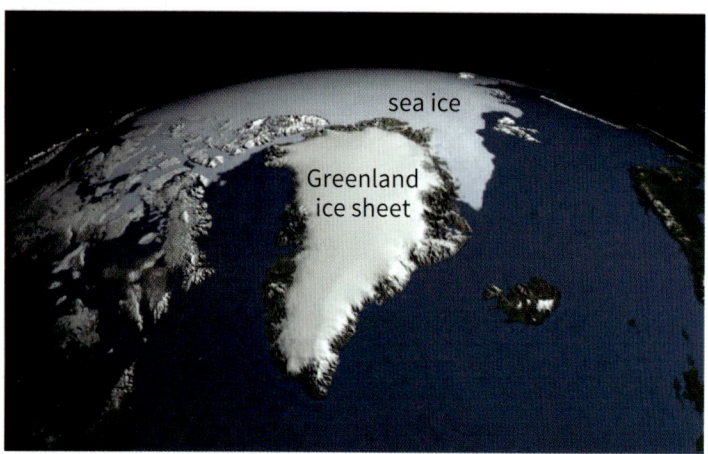

▲ Greenland in winter – see X on map **B**. Its ice sheet loses more ice each year. If it all melts, sea levels will rise by over 7 metres.

▲ When ice melts, the ground warms up, because darker surfaces absorb more sunlight. It's the **albedo effect**.

Climate change

It's a crisis!

Climate change is having an impact around the world – and it already means disaster for many people. Look at these examples.

Sea levels are rising, because ice sheets are melting, and water expands as it warms. So low-lying coasts are having more **floods**.

Overall there's more rain, which also means more floods. But some regions are getting less rain than before. They suffer **drought**.

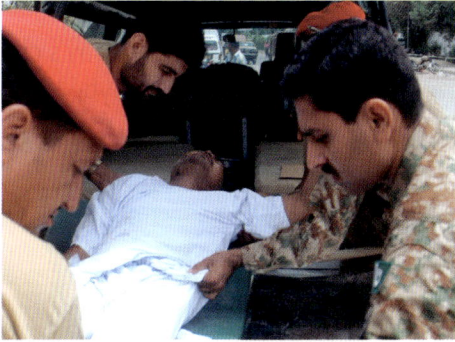

Heatwaves are becoming more frequent and severe. They can kill crops, wildlife, and people – and especially babies and old people.

Along with heatwaves, **wildfires** are becoming more frequent too. These destroy homes, crops and trees, and kill off wildlife.

Crop yields are falling in some regions because the climate no longer suits these crops. People are going hungry.

Pests are spreading to places that were once too cold for them. Mosquitoes bring malaria and other diseases. Crop pests kill crops.

Earth is still warming. The impact of climate change is growing. Some places have already become too difficult to live in. People are moving. It's a crisis!

Your turn

1. Look at **A**.
   a. What do the stripes show?
   b. Write a paragraph to describe how Earth's average temperature changed across the period 1850 – 2018. Include the word *trend* in your paragraph.

2. **B** shows how temperatures had changed from the average for 1951 – 1980.
   a. i. Which *hemisphere* has warmed more? (Glossary?)
      ii. Which part of this has warmed the most?
   b. From **B**, identify two countries which had warmed by between 1.5 and 2 °C. (Page 141 may help.)

3. The ice sheets are melting much more slowly in Antarctica than in Greenland. Use **B** to explain why.

4. The Greenland ice sheets are melting faster than was first predicted, because of the *lobeda ceftef*.
   Unjumble the term in italics, and then explain it.

5. Which effects of climate change might make people move?

6. True or false? Give evidence. (You may need the glossary.)
   A Climate change is having social consequences.
   B Climate change has no economic consequences.
   C Climate change has no environmental impact.

6.4 This time … is it us?

 Are we humans causing today's climate change? Most scientists say we are. Find out more here.

Yes, it's us!

Most scientists agree that today's climate change is not natural. They say we humans are the main driver. Look:

- Temperatures are rising faster than at any time in the last 2000 years.
- Natural changes – such as the changes in Earth's movements – are going on as usual. But they do not explain this temperature rise.
- The temperature rise echoes the rise in greenhouse gases in the atmosphere.
- And we are causing the rise in greenhouse gases.

▲ *Greta Thunberg from Sweden. At age 15 she inspired a wave of protests by young people around the world, about the failure of adults to tackle climate change.*

What are greenhouse gases?

Greenhouse gases are gases that act like a blanket in the atmosphere, trapping heat around Earth. To see how they work, follow the numbers in **A**.

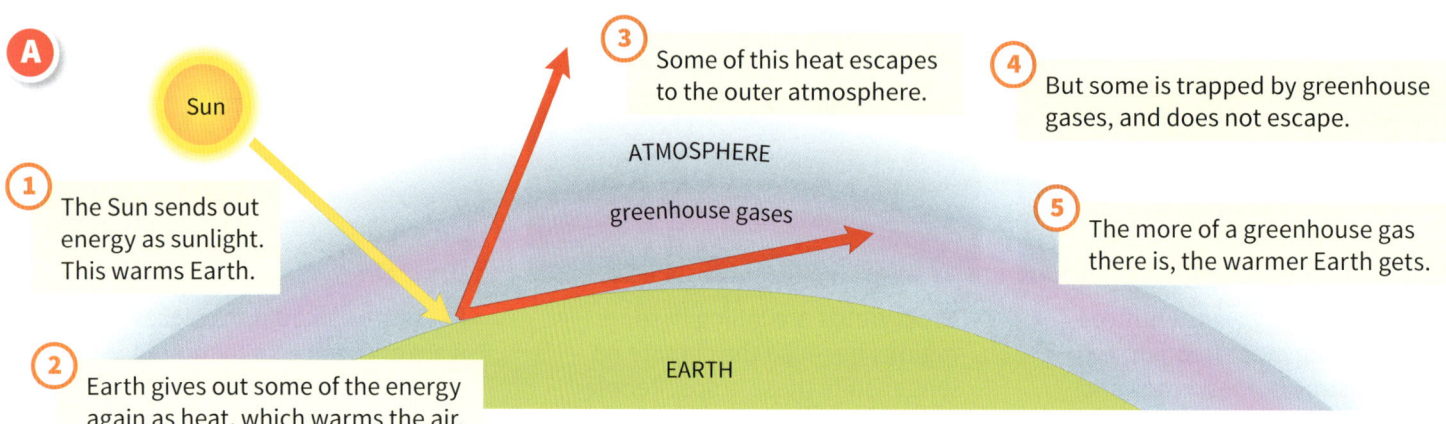

A

1. The Sun sends out energy as sunlight. This warms Earth.
2. Earth gives out some of the energy again as heat, which warms the air.
3. Some of this heat escapes to the outer atmosphere.
4. But some is trapped by greenhouse gases, and does not escape.
5. The more of a greenhouse gas there is, the warmer Earth gets.

We're pumping them out!

We need greenhouse gases in the atmosphere. Without them, all heat would escape from Earth. We'd freeze!

But now the levels are too high. Because we are adding billions of tonnes of them to the atmosphere each year. And this is making Earth warmer.

The main ones are **carbon dioxide** and **methane**. Scientists can tell how much their levels have risen by examining the air trapped in ice cores. (Page 94.)

Look at **B**. It compares the global temperature change (relative to the average for 1891 – 1910) with carbon dioxide levels, for the years 1880 to 2017.

They follow a similar path. The trend for both is … **up**.

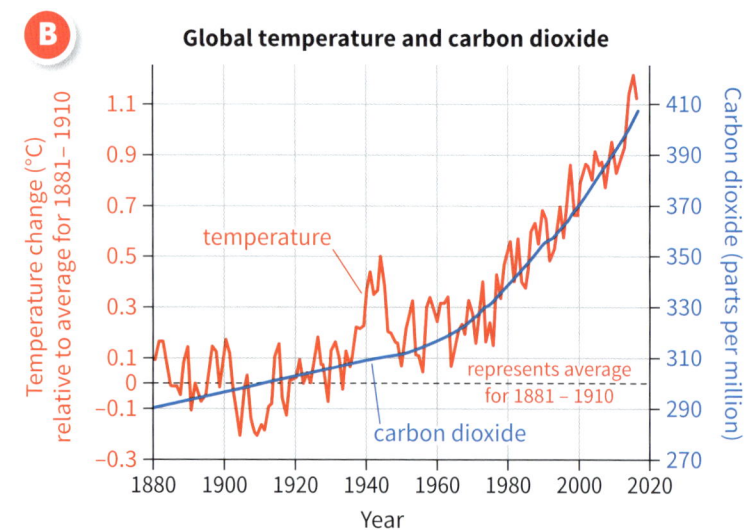

98

Climate change

More about carbon dioxide and methane

Carbon dioxide, CO_2 This is the main culprit in global warming.

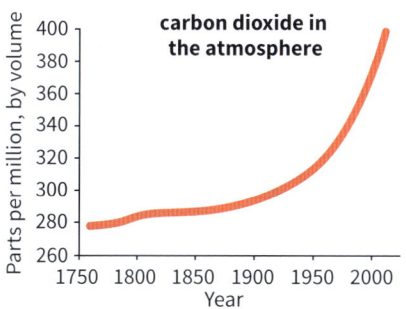

We breathe out carbon dioxide. Vegetation takes it in. It used to be in balance, in the atmosphere. But we add extra, when we burn …

… coal, oil, gas (which is methane), and petrol. We also cut down trees and build on land. So there's less vegetation to absorb it.

The amount in the atmosphere has been rising since the start of the Industrial Revolution, when we began using coal heavily.

Methane, CH_4 Look where this comes from.

Cows, sheep, goats, camels – and other animals that 'chew the cud' – belch out methane. (We think dinosaurs did too!)

Methane is also given off from swamps, and paddy fields, and landfill sites. Some escapes from oil and gas wells.

Every year, we raise more animals, grow more rice, extract more oil and gas, and bury more rubbish. So methane levels are rising too.

Methane is a powerful warming gas – around 25 times more powerful than carbon dioxide. But CO_2 is the main culprit because we put far more of it into the atmosphere. (Compare the labels on the *y* axes on the graphs.)

Your turn

1. a Define the term *greenhouse gas*.
    b Explain how greenhouse gases help to warm Earth.
2. Copy and complete these statements.
    a We'd die without greenhouse gases, because …
    b Greenhouse gases can harm us, because …
    c The two main greenhouse gases are …
    d Carbon dioxide occurs naturally in the atmosphere, but now …
    e The amount of carbon dioxide in the atmosphere has increased since the Industrial Revolution, because …
3. Describe any patterns you notice in graph **B** on page 98.
4. Wood contains carbon. When it burns, carbon dioxide forms.
    a Predict the effect of forest fires on global warming.
    b Planting trees helps to slow global warming. Why?
5. Describe three ways in which we humans help to increase the level of methane in the atmosphere.
6. Look at what these two think. Write serious replies.

6.5 Local actions, global effects

When we emit greenhouse gases, it affects the whole world. Here we concentrate on the main culprit: carbon dioxide.

The trouble is …
When we add greenhouse gases to the atmosphere, we affect everywhere.

For example, carbon dioxide forms when we burn **fossil fuels** – coal, oil and gas – in power stations (for electricity) and factory furnaces.

We burn gas for heating and cooking. (It gives less carbon dioxide per unit of heat than oil or coal. Coal gives the most.)

Most vehicles burn petrol or diesel. Planes burn kerosene (from oil). Ships burn fuel oil. So carbon dioxide streams from exhausts.

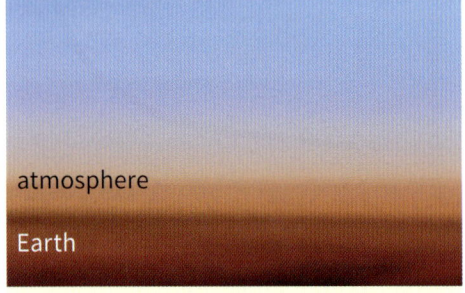

But the gas does not stay where it was produced. It is carried around in the atmosphere, causing warming everywhere.

So our carbon dioxide emissions affect living things everywhere. Even in the ocean. It's an example of **local actions, global effects**.

Many poorer countries contribute very little to the carbon dioxide in the atmosphere. But they still suffer the impacts of climate change.

Carbon dioxide emissions per person

Look at **A**. It shows **emissions** of carbon dioxide *per person* for eight countries for two years: 2000 and 2017. In tonnes!

(To get these figures, you divide a country's total emissions of the gas, from burning fossil fuels, by its population.)

Look at the note about Chad. It is a poor country. It uses oil to generate electricity. But in 2017, only 11% of its people had electricity. Very few had cars.

Look at Qatar. This wealthy desert country has lots of gas, which is its main fuel.

By 2017, some countries had already managed to reduce their emissions.

Look at the UK. What do you notice about its emissions?

A Carbon dioxide emissions per person (tonnes)

*Chad's emissions were so low that they rounded off to zero.

Climate change

The bad news is ...

It's over 100 years since scientists first noticed a link between carbon dioxide and Earth's temperature.

In 1990, scientists warned that the world would keep warming if we did not cut greenhouse gas emissions.

Countries agreed to reduce emissions in treaties called the **Kyoto Protocol** (1997) and **Paris Agreement** (2016).

But by 2018, world emissions were still rising. Look at **B**.

If we carry on ...

Suppose the world's total carbon dioxide emissions keep rising, as in **B**. Scientists predict that the global temperature will rise a further 2 – 3 °C by 2100.

This would spell **disaster** for much life on Earth.

Scientists say the best we can hope for is to **limit** the rise to **1.5 °C** above the average for the years 1850 – 1900.

It has already risen by around **1 °C** since then.
So ... **0.5 °C** to go.

Can we stay within this limit? And how?
Explore these questions in Unit 6.6.

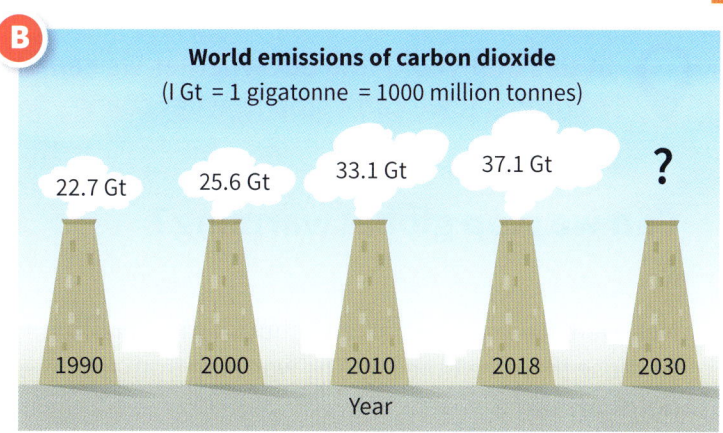

▲ A positive message from climate protesters.

Your turn

1 We all burn fossil fuels, directly or indirectly.
   a Name the fossil fuels.
   b Explain how they are linked to global warming.
   c Give one example where people burn fossil fuels *directly*:
      i at home
      ii when travelling
   d Explain why this may cause fossil fuels to be burned *indirectly*:
      i switching a light on
      ii charging a mobile

2 Bar chart **A** shows emissions of carbon dioxide *per person*.
   a Define *emissions*. (Glossary?)
   b Comment on the emissions per person for Qatar.
   c Chad is in Africa. In 2017, people in Chad had about £10 a week to live on, on average. For *everything*! Suggest three reasons why the carbon dioxide emissions per person for Chad were too low to show up.

3 Chad is getting hotter and its rainfall is decreasing.
   a Explain why Chad is getting hotter, even though its own carbon dioxide emissions are very low.
   b Nearly 90 % of the people in Chad live by farming. This means Chad may suffer greatly as its climate changes. Explain why.

4 Look again at the carbon dioxide emissions in **A**.
   a How did emissions per person change for China between 2000 and 2017?
   b From what you have learned so far, do you think China grew wealthier, or poorer, over that period? Explain.
   c Identify the other countries in **A** where emissions per person grew in the period 2000 – 2017.
   d In 2017, China's emissions of carbon dioxide per person were lower than Japan's. *But in total, China emitted 8 times more carbon dioxide that year than Japan did.* See if you explain the fact in italics.

5 Which countries in **A** managed to *reduce* emissions in the period 2000 – 2017?

6 Explain why the trend shown in **B** is bad news for:
   a humans
   b polar bears, which live in the Arctic

7 **C** shows a protest in London. Explain why people are carrying photos of people in other countries.

8 Suggest three things *you* could do in your everyday life, to help reduce carbon dioxide emissions. Think carefully!

6.6 What can we do?

In this unit you'll find out whether we can stop global warming. If we can't stop it, what can we do?

Can we stop global warming?

No.

Not even if we stopped all greenhouse gas emissions this minute. Because there's a time lag. The warming from recent emissions will not show up for years.

But we can't do nothing. And our choice is stark:
- The faster we cut emissions now, the better our chances of surviving and coping in the future.
- The more slowly we act now, the greater the crisis ahead.

Our best hope is to limit the temperature rise by the end of this century to 1.5 °C above the average for 1850 – 1900.

To do this, we must cut emissions of carbon dioxide heavily by 2030. And be **carbon neutral** by 2050 at the very latest.

At the same time, we need to prepare for a warmer future. It's on the way. We have to adapt.

Most countries have set targets to limit carbon dioxide emissions. They must stick to them. It's not easy, because the world is addicted to fossil fuels. But there is no option.

Everyone can do something. Even if it's only switching the light off in an empty room. It all adds up.

▲ In the tundra, global warming is causing the deeply frozen soil below ground to melt. This releases the greenhouse gas methane. And houses subside!

▲ Sea levels are predicted to rise by 1 m by 2100, if we don't cut emissions. This is Male, chief island of the Maldives in the Indian Ocean. A 1-metre rise would submerge it.

Ways to cut emissions

Around the world, people are pushing on with projects to cut greenhouse gas emissions. For example:

- wind farms, which generate electricity from the wind
- solar farms, which generate electricity from sunlight
- wave and tidal power plants, which generate electricity from the movement of the waves and tides
- electric cars and other vehicles that run on batteries (ideally charged by electricity that's not from power stations which burn fossil fuels)
- artificial meat 'grown' in factories, which tastes very like real meat

Over time, we'll develop other solutions too.

But the key thing is this: we must cut emissions **fast.** And each of us can play a part. Get cutting!

▲ Walney Wind Farm, off Cumbria (on England's west coast). The UK is one of the world's best locations for wind power.

Climate change

Some strategies

Here are some strategies to tackle the climate crisis. Some are about cutting emissions. But not all! What do you think of them?

A Find ways to remove carbon dioxide from the air.
B Reduce the amount of carbon dioxide that's produced.
C Reduce the amount of methane that's produced.
D Stop some sunlight from reaching Earth.
E Help people to adapt to changing climates.

Your turn

1 a Is it possible to keep Earth's average temperature to what it is today? Explain your answer.
   b If we don't cut emissions urgently, what will the outcome be, for your children and grandchildren?
   c Ideally the world will be carbon neutral by 2050. Explain what *carbon neutral* means. (Glossary?)

2 List four ways to generate electricity that do not produce carbon dioxide.

3 Explain how 'growing' artificial meat in factories could help us to reduce greenhouse gas emissions.

4 Look at actions 1 – 16 over on the right, to tackle the climate crisis.
   Match each action to one of the strategies **A – E** above, using a table like this. (One number has been filled in for you.)

Strategy	Action
A	
B	7
C	
D	
E	

5 Look again at the list of actions 1 – 16.
   a Choose any *two* which you think depend most on:
      i   scientists
      ii the government
      Explain your choice.
   b People will protest about some of those actions. Which two might they protest about most? Explain each choice.
   c Which actions, if any, might turn out to be harmful for Earth, and difficult to reverse? Give reasons.

Actions to tackle the climate crisis

1 Build more solar farms and wind farms.
2 Put big taxes on air travel.
3 Build new homes that are able to keep out floods.
4 Breed new crops that need less water.
5 Turn off all street lights at midnight.
6 Find a way to capture the carbon dioxide from power stations and store it underground.
7 Ban the use of gas for cooking and for heating homes.
8 Spray fine mists (aerosols) into space, to reflect sunlight away. (They act like volcanic gases.)
9 Pay countries to protect their rainforests.
10 Build homes that do not need heating in winter, or cooling in summer.
11 Allow homes to have electricity for only 6 hours a day.
12 Put solar panels on the roofs of all buildings.
13 Plant more forests.
14 Double the price of meat.
15 Make the roofs of all buildings white.
16 Ban power stations that use fossil fuels.

6 Which actions above will affect the design of future homes?

7 You are in charge of tackling the climate crisis. Which *three* actions above will you choose first? Explain your choice.

8 Design a poster asking people to do what they can to help cut greenhouse gas emissions. Choose a good slogan for it.

6 Climate change

How much have you learned about climate change? Let's see.

1 **A** shows a car getting a fill of petrol, in the UK.
   a Petrol is made from a fossil fuel. Name the fossil fuel.
   b i Name the greenhouse gas that forms when petrol burns in the car engine.
      ii State another source of this greenhouse gas.
   c Kutubdia is a low-lying island off Bangladesh. It has been partly drowned by rising sea levels. People have lost homes and livelihoods.
      Explain the link between photos **A** and **C**.
      You could draw a flow chart, or write bullet points.

2 a The car in **B** runs on electricity. It is better if the electricity does *not* come from power stations like the one in the first photo on page 100. Explain why.
   b Name four ways to generate electricity, that do not produce carbon dioxide.
   c The sources of electricity that you named in **b** are described as *renewable*. Explain why. (Glossary?)

3 The UK is working hard to cut carbon dioxide emissions. **D** shows the change in emissions from 1990 to 2018.

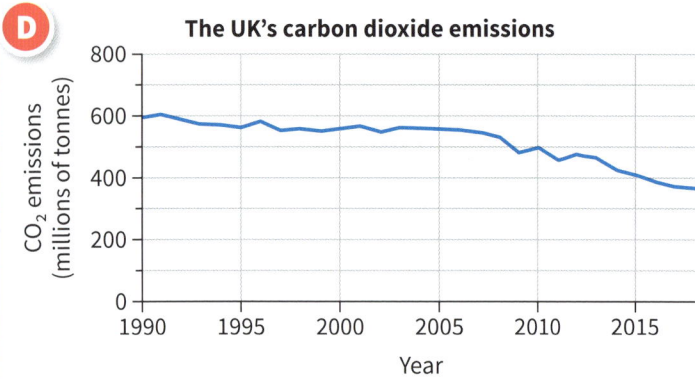

   a About how much carbon dioxide was emitted in:
      i 1990?   ii 2015?
   b Do you think the UK was successful at cutting CO_2 emissions over that period? Justify your answer.
   c These steps were taken in the UK over that period. Explain how each helped to reduce CO_2 emissions.
      i   Many windfarms were built, including off the coast.
      ii Three coal-burning power stations were closed.
      iii Many people fitted solar panels to their roofs.
      iv The government gave people grants to insulate their homes, to reduce heat loss in winter.
   d As well as cutting emissions, we could remove carbon dioxide from the air. State one natural way to do this.
   e Name one other greenhouse gas.
   f Moving to a more plant-based diet can help to reduce greenhouse gas emissions. Explain why.

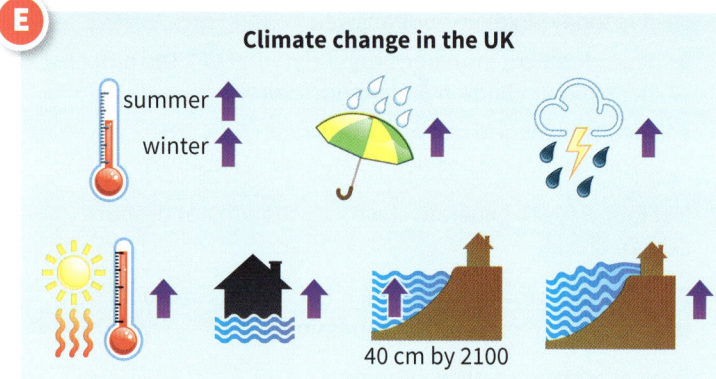

4 **E** shows how the UK's climate is changing.
   a Using **E** to help you, write a paragraph describing how the UK's climate is changing.
   b We must take steps to adapt to the changes in **E**.
      i   Define *adapt*.
      ii One suggestion is to put shutters on all windows. Explain how this would help us.
      iii Suggest five other things we could do, to adapt to the changes shown in E.

5 *Countries must work together to prevent a climate catastrophe.*
   To what extent do you agree with this statement?
   Give reasons to support your answer.

Asia

7.1 What and where is Asia?

 This unit will remind you where Asia is – and you can compare it with the other continents.

Did you know?
- The big land mass that's shared by Europe and Asia is called Eurasia.

Asia: a continent

Asia is one of the world's seven continents. Look at map **A**.

A [World map showing the seven continents with points P, Q, R, S, X, Y marked]

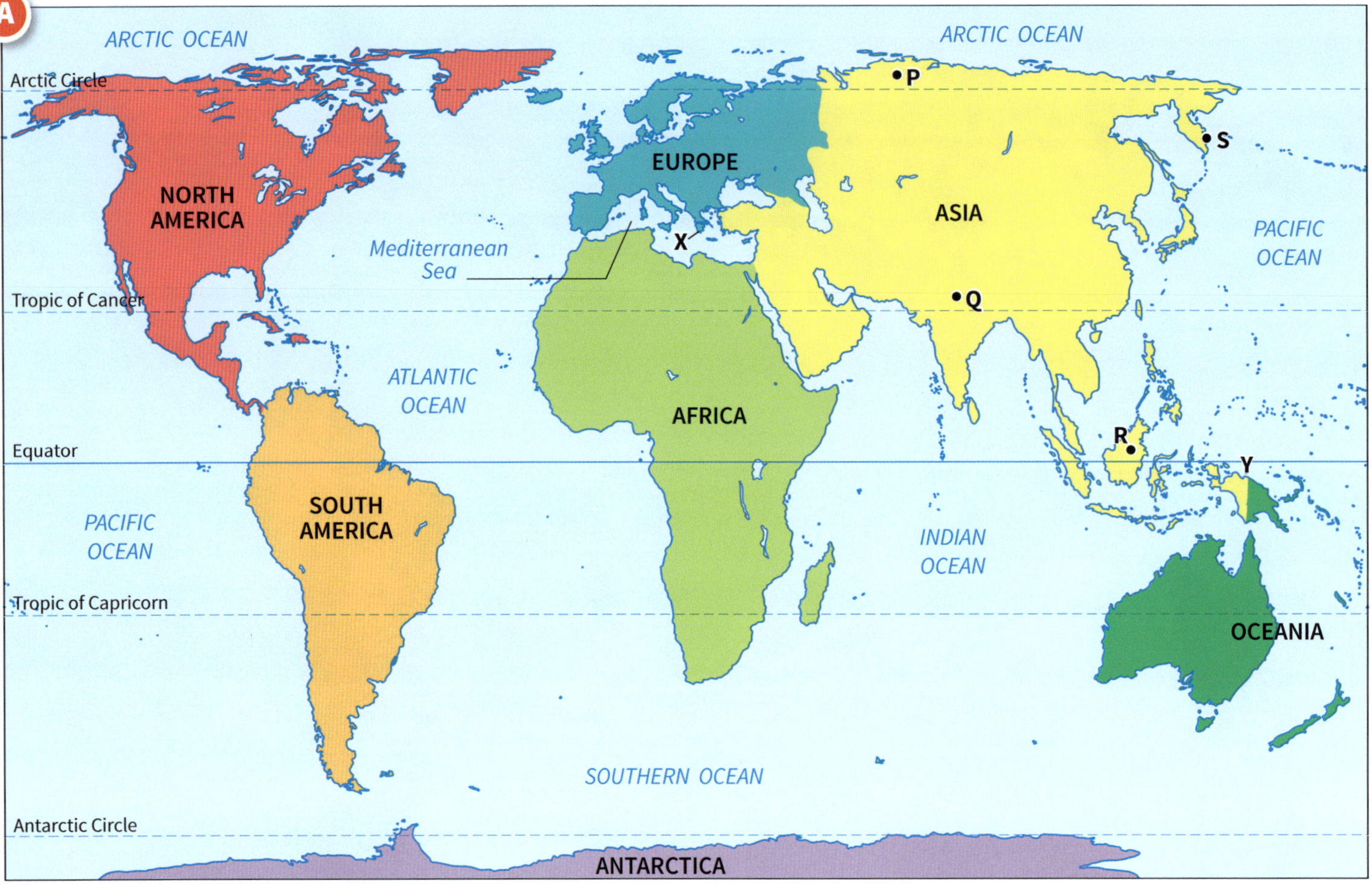

Where are Asia's borders?

With Europe

Asia is joined to Europe. So why do we call it a separate continent? That began with the Ancient Greeks, around 2500 years ago. (Greece is at X on map **A**.) They didn't know much about the world to the east of Greece. So they called it *Asia*, from a word that means *east*.

Since then, people have argued about Asia's border with Europe. Today, most people accept the one shown on map **B**. It cuts through some countries. For example most of Russia lies in Asia – but most of its people live in Europe!

With Oceania

Asia's border with Oceania mostly wiggles between islands. But look at New Guinea, the island at Y on map **A**. Half of it is in Asia, and belongs to Indonesia. The other half is in Oceania, and is called Papua New Guinea.

Asia

The largest continent

Asia is the world's largest continent, for both area and population.
Look at these two tables.

What if …
… the UK was a country in Asia?

The continents by land area

Continent	millions of square km
Asia	44.6
Africa	30.1
North America	24.5
South America	17.8
Antarctica	13.2
Europe	9.9
Oceania	8.1

The continents by population

Continent	millions of people
Asia	4545 (or 4.545 billion)
Africa	1287 (or 1.287 billion)
Europe	743
North America	588
South America	428
Oceania	41
Antarctica	people only visit

Your turn

1. This graph compares the **areas** of the seven continents.

   The continents by land area

   Using the graph, decide whether each statement below is true, or false. If false, rewrite it to make it true.
   - A Together, Africa and Europe would fit into Asia.
   - B Together, North and South America would fit into Asia.
   - C Asia is over five times the size of Europe.

2. This pie chart compares the **populations** of the continents.

   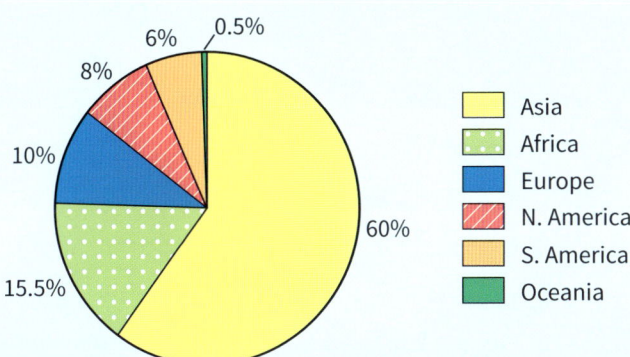

   Share of Earth's population

   True or false? If false, write a correct statement.
   - A More than half of the world's population lives in Asia.
   - B Asia has over five times more people than Europe has.
   - C There are over ten times more people in Asia than in North America.

3. Look at the blue lines on map **A**.
   - a Describe where the Equator passes through Asia.
   - b Do any other major lines of latitude pass through Asia? If yes, name them.
   - c Look at the places marked P, Q, and R on the map.
     - i One is very cold, with long dark winters. Which one?
     - ii One is hot and wet all year, and still has some tropical rainforest left. Which one?

     Each time, explain your choice.

4. Look at the oceans on map **A**.
   - a Write a paragraph about the oceans which border Asia. You must include their names, and these words:

     coast   north   south   east
   - b Imagine you start at S on map **A** and sail east across the Pacific Ocean. Where will you end up?

5. Look at the border between Asia and Europe, on map **B**. It makes use of physical features. For example, it follows the Ural Mountains.

   Write a paragraph to describe the border's route, from north to south. Include all these terms:

   Arctic Ocean   Black Sea   Caspian Sea
   Mediterranean Sea   Ural Mountains

6. Asia is a continent.
   - a First, explain the difference between a *country* and a *continent*. (Glossary?)
   - b Now, a challenge. Name as many Asian countries as you can, *without looking at a map*. (There are 49.) Try to come up with at least five. No peeking!

7. It's time to start a spider map for Asia.
   - Use two pages, so that you have plenty of room.
   - Mark in facts you know already.
   - Try to group information under headings. (For example *Population, Oceans and seas, Mountains …*)

   You can add to your spider map as you work through this topic. It will become your summary for Asia.

7.2 Asia's countries and regions

Asia has 49 countries. Find out more about them here.

The countries and their capitals

Some Asian countries are huge. Some are tiny. Map **A** shows their capitals too.

B Compare!

*China claims that Taiwan is a province of China, and not an independent country.

Key
- CHINA — country names are shown like this
- ■ capital cities
- — country boundary
- ---- disputed country boundary
- ······ continental boundary

Asia's regions

Asia's countries are grouped into regions. Map **C** shows the regions.

Look at West Asia. It is usually called *the Middle East*.

▲ The Heart of Asia monument marks the centre of Asia – near the town of Urumqi in China. (See **X** on the map on page 108.)

Your turn

1 Find these countries on map **A**, and name them.
   a It is sandwiched between Russia and China.
   b It lies to the east of China, and its name starts with J.
   c It borders the Arctic Ocean.
   d Sri Lanka lies off its coast.
   e The Equator passes through this country of many islands.
   f It shares borders with China, Laos, and Thailand.
   g This small island country lies at the tip of another country, near the Equator. It has one of the busiest ports in the world. Its name has 9 letters, and ends in *e*.
   h These two countries, near Japan, were once a single country. You can guess this from their names.
   i Its name has four letters, and ends in *q*.

2 Now see how many countries you can find, beginning with:
   a M   b P   c I   d T

3 Find each capital city below on map **A**. Then write down the capital, and its country.
   a Riyadh   b Jakarta   c New Delhi
   d Kabul   e Hanoi   f Beijing
   g Pyongyang   h Kuala Lumpur   i Moscow
   j Manila   k Islamabad   l Ulan Bator

4 Of Asia's 49 countries, 27 are smaller than the UK in area. Map **B** shows the UK at the same scale as map **A**.
   a Pick out four Asian countries that are bigger than the UK.
   b Now pick out four you think are smaller than the UK.

5 Map **C** shows Asia's regions. Name countries as follows:
   a five in Central Asia, all with names ending in -*stan*.
   b four in South Asia
   c the only country in North Asia
   d five in West Asia (the Middle East)
   e five in East Asia
   f five in Southeast Asia

6 Describe the location of Pakistan. Mention its region, its neighbours, an ocean, and the Tropic of Cancer!

7 The *Middle East* was given that name by Europeans. Which name do you think suits it better: *the Middle East* or *West Asia*? Give your reasons.

8 a The border between Asia and Europe is shown on map **A**. Name two countries it passes through.
   b In which Asian country would you find polar bears?
   c Identify all the Asian countries that lie *completely* within the tropics. (Page 141 may help.)

7.3 What's Asia like?

This unit will give you an overview of Asia today.

Why ...
... are the written letters so different, for some languages?

What if ...
... Asia and Europe were just one continent?

The most diverse continent of all

Asia is the largest continent. It has more people, more cultures, and a bigger range of climates and environments, than any other continent.

Asia's people

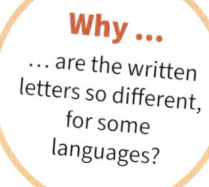

- 60% Asia
- 40% Rest of the world

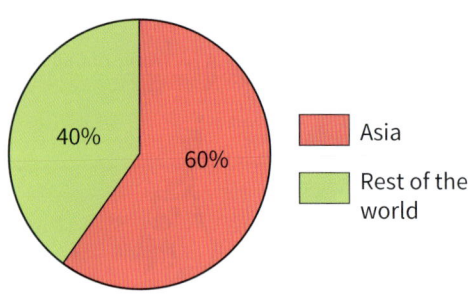

Asia has around 4.6 billion people. Roughly 60 out of every 100 people on Earth live in Asia. (That's 60%.)

Over half (2.8 billion) are in just two countries, China and India – the world's two most populous countries.

Across Asia there are thousands of different ethnic groups. India alone has more than two thousand.

Happy birthday ... in

Arabic	عيد ميلاد سعيد
Chinese	生日快乐
Russian	С днём рождения
Thai	สุขสันต์วันเกิด
Urdu	سالگرہ مبارک ہو

There are thousands of languages too. (Some are not very widely used.) Here's *Happy birthday* in a few.

Many religions are practised in Asia. The top three (most common first) are Islam, Hinduism, and Buddhism.

About 53% of Asia's population lives in rural areas – mostly as farmers. (The figure for Africa is about 60%.)

But Asia has enormous cities too. 8 of the world's 10 largest cities are Asian. This is Shanghai, in China.

Around 10% of Asia's population lives in poverty, on less than £2.50 a day. But the poverty rate is decreasing.

There is enormous wealth too. The number of super-rich people in Asia is growing very fast.

110

Asia's economy

How are Asia's regions doing? Follow the numbers to find out …

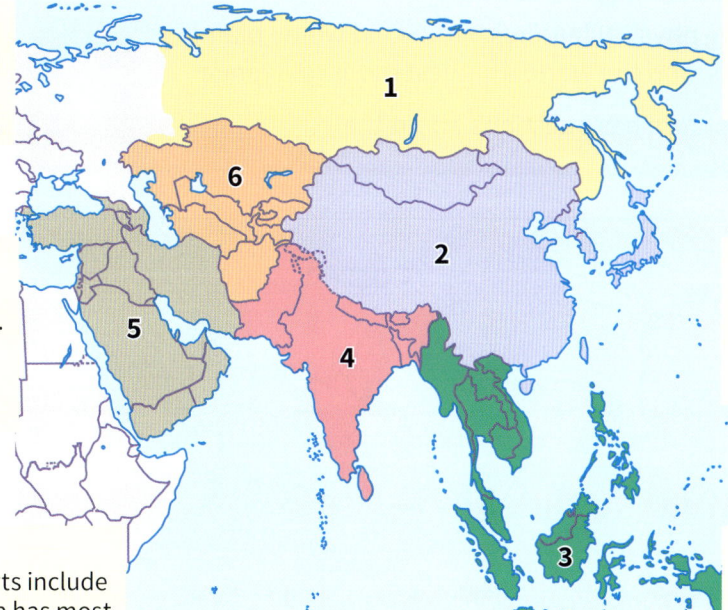

6 Central Asia
Not so well off. Kazakhstan has the strongest economy in this region, thanks to its oil, gas, and other minerals.

5 The Middle East
It has a big share of the world's oil and gas reserves. This has made some countries very wealthy. Qatar, for example.

4 South Asia
Asia's poorest region. Exports include textiles and foodstuffs. India has most of the industry, and runs IT services, and call centres, for other countries.

1 North Asia
Russia is doing okay. It sells oil and gas to other countries. It exports them by pipeline.

2 East Asia
The main industrial region. You're sure to own things that were made here – in China, or Japan, or South Korea, or Taiwan.

3 Southeast Asia
These countries depend mainly on agriculture. They export rice and other foodstuffs, and rubber. But industries are developing too.

Even within a region, a country can be much richer or poorer than the country next door. And even if people in a country are well off *on average*, some are always much richer or poorer than others.

Your turn

1. Name the two most populous countries in Asia.
2. Pick out what you think are the three most interesting facts about Asia from page 110.
3. Look at the map above. Which of Asia's regions:
   a. is the main industrial region?
   b. has a large share of the world's oil deposits?
   c. is poorest, overall?
4. The table on the right shows data for a recent year. *GNI per person (PPP)* is a way to measure the economy. It gives you an idea of how well off people are, on average.
   a. What does *GNI per person (PPP)* mean? (Glossary.)
   b. i For which country in the table is it highest?
      ii Why is that country so well off? (Look above.)
      iii About how many times better off are people there, on average, than people in the UK?
   c. Now choose another country *in the same region as the country in b*, where people are much poorer. (Page 141?)

Country	GNI per person (PPP) in dollars
Afghanistan	1950
Bangladesh	4360
China	18 210
India	7760
Indonesia	13 060
Japan	43 350
Qatar	126 600
Russia	27 150
Yemen	2570
UK	46 240

5. Which country in the table above is closest to the UK, for GNI per person (PPP)?
6. *Asia is a continent of great contrasts*. Give evidence to support the statement in italics, using what you learned in this unit. Write at least half a page!

7.4 What are Asia's main physical features?

Here you can learn about Asia's key physical features – and which countries they lie in.

Mountains, rivers, deserts, glaciers …

Asia has a huge range of amazing physical features. From cold windy plains to scorching hot deserts, glaciers, and the world's highest mountain range.

Map **A** shows the main ones. *Your turn* has questions about them. Get ready to be a map detective!

Did you know?
- The Himalayas are growing taller by about 6 cm a year.

A

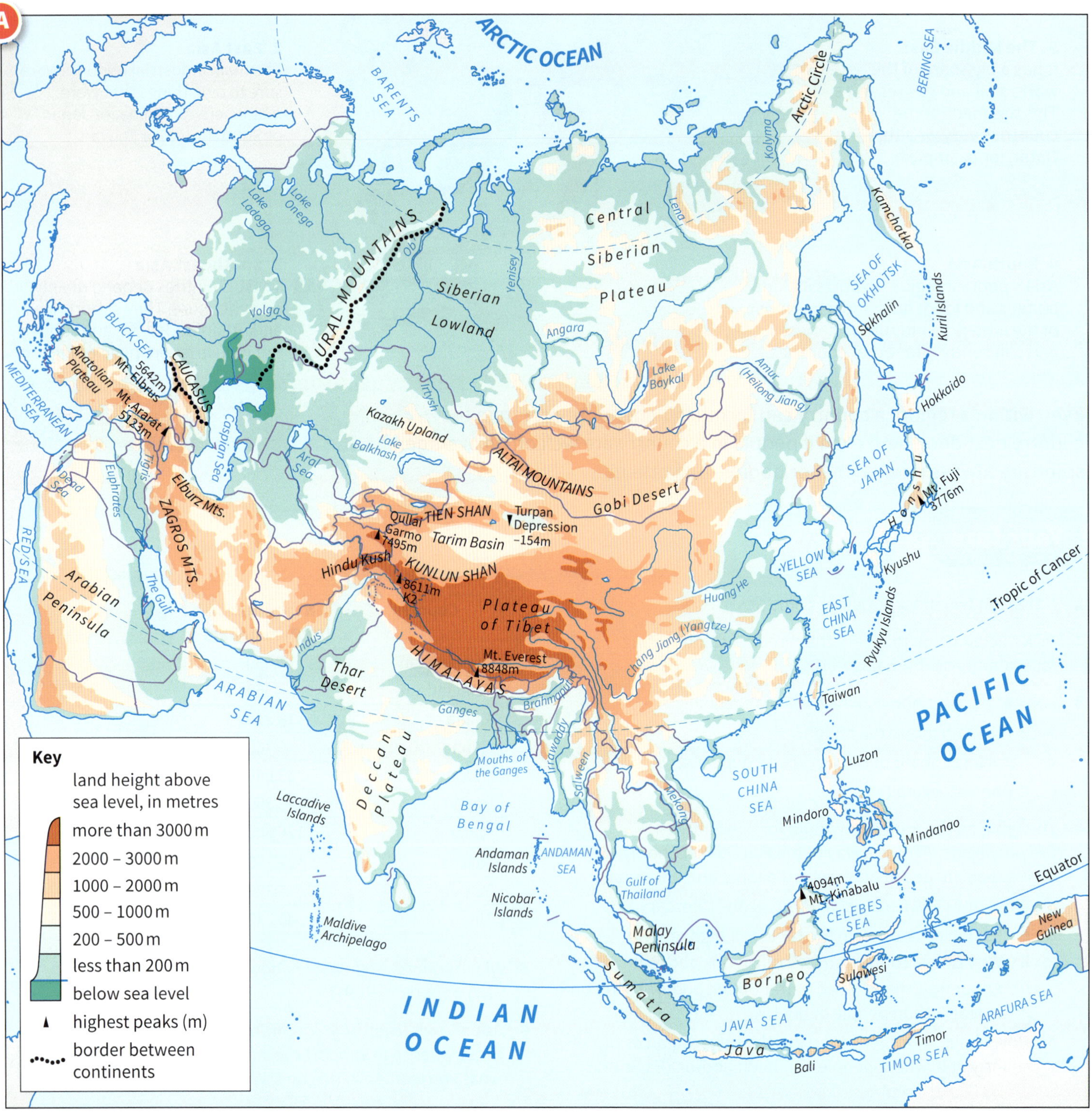

Your turn

Asia

The map on page 141 will also help you to answer these questions.

1. First, study the satellite image of Asia on page 105. Then, with its help, describe Asia's physical geography. Write a paragraph. You could include these terms:
   mountainous   flat   vegetation   coastline   oceans   islands

2. Photo **B** was taken on the Plateau of Tibet. This plateau is about 4 km above sea level, on average. It is sometimes called 'the roof of the world'. Find it on map **A**.

   a Define the term *plateau*. (Glossary?)
   b Most of the Plateau of Tibet lies in one country. Which one?

3. Asia has other big plateaus too – but not so high. Name the plateau marked on the map that lies in:
   a India   b Turkey   c Russia

4. Eight big rivers rise in the Himalayas and Plateau of Tibet.
   a Name four of them, and the sea or bay each flows into.
   b Of the eight, which one matches this description?
      i   It's the third longest river in the world – and it flows through only one country.
      ii To Hindus, it's a sacred river. It has many mouths.
      iii Many people in Pakistan depend on this river for water.
      iv This river flows through six countries. One is Cambodia.

5. Asia has the world's highest mountain range. It's shown in **C**.
   a This mountain range is called …?
   b Which countries share it?
   c It contains the world's highest mountain, called …?
   d The world's second highest mountain is also in this range. It's in Pakistan. Name it too.
   e The landform just north of this mountain range is …?

6. Asia has other mountain ranges too. Name:
   a i two that are fully in China   ii two that lie in Iran
   b The Altai Mountains are shared by four countries. Identify these countries. (Hint: C, K, M, R.)

7. Asia has the world's largest lake. No water flows out of this lake. It's a bit salty, so it is usually called a *sea*.
   a Its name begins with *C*. Identify it.
   b Which countries border this sea?
   c The map shows a large river that feeds this sea. Name it.

8. a Identify the oceans that border Asia.
   b Areas of ocean that border land are often called *seas*. Name the sea that lies:
      i   off the north east tip of Russia
      ii between Vietnam and the Philippines
      iii off the coast of Pakistan

9. Now look at the Bay of Bengal.
   a Name three countries that border it.
   b Explain the difference between a *bay* and a *sea*. (Glossary?)

10. a Define *peninsula*.
    b Asia's largest peninsula is shaped like a wellington boot. It is mostly hot desert. Name it, and three countries in it.
    c Which peninsula has the island of Singapore at its tip?

11. The Gobi desert is a large cold desert, about 5.5 times the size of Great Britain. It gets bitterly cold in winter.
    a Why is it cold, not hot?   b Which two countries share it?

12. Look at satellite image **D**.
    a X marks a hot desert shared by India and Pakistan. Name it.
    b Name the river shown as a blue line.
    c i Name the mountain range marked Y.
       ii How can you tell from the satellite image that this mountain range is high?

7.5 Asia's population

Some parts of Asia are crowded, and some are almost empty. Here you can find out more, by exploring a map.

Where is everyone?

Asia is the most populous continent. It has over 4.5 billion people!

The map below shows the **population density**. The deeper the shade, the more people live in that area. And note the symbols for cities too.

Did you know?
- Only ten Asian countries have more people than the UK (not counting Russia, which has most of its population in Europe).

A

Key

Population density
people per square kilometre
- over 100
- 10–100
- 1–10
- under 1
- ⋯⋯ continental boundary

Major cities
population in millions
- ▢ over 3
- ■ 1–3
- • 0.5–1
- · 0.1–0.5

Asia

▲ A village with rice fields on the Indonesian island of Java. Indonesia has over 17 000 islands, and is Asia's third most populous country.

▲ Singapore, Asia's second smallest country by area. It is an island. 90% of it is covered by the city, also called Singapore.

Your turn

1. a Define *population density*. Glossary?
   b Look at the places in **B** and **C**. Which one has a higher population density? Explain your choice.

2. From map **A**, decide whether each statement is true, or false. If false, rewrite it to make it true.
   a Around half of Asia is only lightly populated.
   b Most people live in the northern half of Asia.
   c Overall, Russia is more densely populated than China.
   d Overall, the Asian part of Russia is more densely populated than the European part.
   e Most of India's big cities (with at least one million people) are in the southern half of India.
   f Most of China's major cities are in the middle of China.

3. Find the four places marked W, X, Y and Z on the map.
   a Which has fewer people, W or Y? Explain why.
   b Compare the population densities at Z and Y, and suggest a reason for the difference. (*Compare*, page 137?)
   c Now match each place W – Z to a country. (Page 141.)

4. Look at bar graph **D**. It shows the ten most populous Asian countries, and the UK for comparison. (Russia isn't included, since most of its people live in Europe.)
   a Which two countries have most people, by far?
   b Which has a larger population, China or India?
   c The population of China is about …?
      Give your answer in: i millions   ii billions
   d Compare population sizes for Iran, Japan, and the UK.

5. Look again at **D**. How many more people are there …
   a in China than in the UK?
      i about 52 times more   ii about 21 times more
   b in India than in Pakistan?
      i about 7 times more   ii about 17 times more

6. Photo **C** shows Singapore, one of Asia's smallest countries. It is very wealthy. It is heavily built up. It has an area of about 700 square km. (So it's about six times the size of Manchester.) It has a population of about 5 600 000.
   a Calculate the population density of Singapore. Give your answer to the nearest whole number, with the correct unit.
   b What problems might this population density cause? Suggest as many as you can.

7. Look at photos **B** and **C**. Suggest three ways in which life may be very different for the people living in those places.

8. China has a very large population. For any government, in what ways might a very large population be:
   a a problem?   b an advantage?
   Think of as many ways as you can. Show each answer in any way you wish. Spider map? List? Annotated drawings?

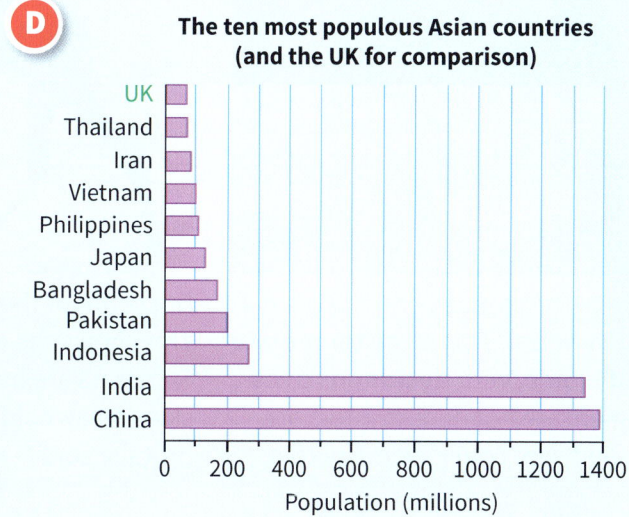

7.6 Asia's biomes

Because of its size, Asia has a wide range of biomes. Find out more here.

Asia's biomes

A **biome** is a large region with its own distinct climate, plants, and animals. The **climate** dictates what a biome is like.

Asia stretches from above the Arctic Circle to below the Equator. It has mountain ranges too. So that means it has many different climate zones – and biomes. Look at these photos. The fills in the circles match map **A**.

Did you know?
- They say the Abominable Snowman (or Yeti) lives in the Himalayas.
- Tall, hairy, walks on two legs, big feet.

In the far north is the **tundra** biome. It is bitterly cold here, and the ground is deeply frozen. But in summer, the surface thaws. Then ponds form everywhere, and low plants grow.

South of the tundra is the **taiga** biome. It has thick **coniferous** forests with trees such as spruce and fir. Winters are long, and very cold. Summers are short, warm, and damp.

Next, in the middle of the continent, are the **steppes**: plains of grassland. Summers are hot, winters very cold. There are few trees, because there is not enough rain to support them.

Between the steppes and the coast it is much wetter. This is the **temperate forest** biome, with **deciduous** trees. Summers are hot here. Winters are cold, and very cold in some places.

But south of the steppes it is very dry. This shows the **cold desert** biome. Summers are hot, but cloudless skies mean cold nights. Winters are brutally cold (– 40 °C or less).

Further south you'll find more desert. But now it is **hot desert**. It is usually very hot during the day, and cold at night. As in the cold desert, vegetation is sparse because there's so little rain.

Asia

Mountain ranges have the **mountain** biome. The higher you go, the colder it is. After a certain point it's too cold and dry for trees. Go high enough and you'll find glaciers.

Furthest south, in and near the tropics, is the **warm moist forest** biome. The forests include **tropical rainforests**, and **mangrove swamps** like this one. (See the map key.)

Did you know?
- Much of Indonesia's rainforest has been cleared to grow oil palm trees.
- Palm oil is used in bread, biscuits, ice cream, shampoo …

Key for photos and map

1	tundra
2	taiga
3	steppe
4	temperate forest
5	cold desert
6	hot desert
7	mountain
8	warm moist forest
	mangrove swamp

Asia's biomes on the map

This simplified map shows the main biomes. The colours and patterns match the circles on the photos.

As you saw earlier, much of Asia is densely populated – and this has affected the biomes.

For example, 8000 years ago, nearly all of Southeast Asia was covered by forest. Now half has gone – cut down for fuel, or cleared away for farmland, and roads, and settlements.

Asia's tropical rainforests are Earth's oldest, and the richest for biodiversity. But they are vanishing fast.

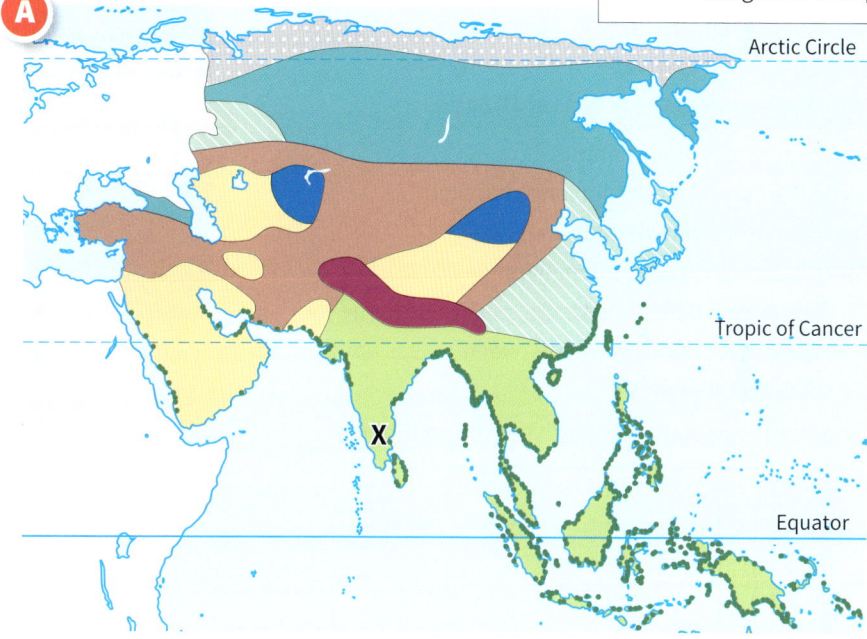

Your turn

1. Define the term *biome*.
2. Asia has a wide range of biomes. Explain why.
3. Name the biome in Asia:
   a. which lies furthest north
   b. where you are most likely to find large flocks of sheep
4. Define each term below. (Glossary?) Then name the biome it is linked to.
   a. temperate   b. tree line   c. coniferous
   d. mangroves   e. permafrost   f. deciduous
5. With the help of the map on page 141, name:
   a. a country with: i tundra   ii dense coniferous forests
   b. two countries with: i hot deserts   ii cold deserts
   c. three countries with mangrove swamps along the coast
6. a. Look at X on map **A**. It lies in the tropics. What kind of warm moist forest might you expect to find at X?
   b. In fact there may not be much forest left at X. Explain why. (The map on page 114 may help.)
7. Discuss the impact of humans on Earth's biomes, using your own ideas *and* what you've learned here.

117

7 Asia

How much have you learned about Asia? Let's see.

1 We are all connected to other continents. Describe two ways in which you are connected to Asia.

2 Start a table with headings like this:

Country	Capital

Then list ten countries which are fully within Asia, and their capitals. (See how many you can list without checking back.)

3 **A** is a smaller version of the satellite image on page 105. Identify the countries where these letters are placed:
   a 3   b 4   c 6   d 1   e 2   f 5

4 This question is also about **A**. You'll need the map on page 112.
   a Define these terms:
      i peninsula   ii sea   iii ocean   iv plateau
   b Name the big peninsula that contains country 1.
   c There's a peninsula at 15. Name it.
   d State the name of the large feature at 7.
   e Name the sea at 8.
   f Name the sea at 9, which is really a salty lake.
   g State the names of the oceans at 10 and 11.

5 **A** shows that Asia is mountainous.
   a Name the curved mountain range labelled 12.
   b Give the names of two mountains in 12.
   c There is a much lower mountain range at 13.
      i This range is called the U_____ . Give its name.
      ii It forms part of the border between _____ and _____. State the missing words.

6 This shows the population of Earth's seven continents.

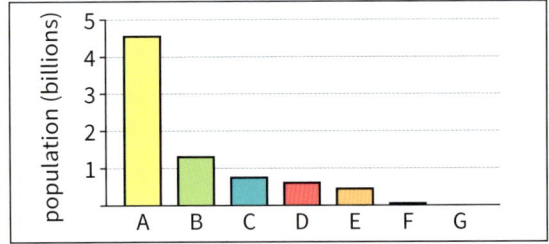

   a Match each bar to its continent. (If you get stuck, the pie chart on page 107 may help.) Answer like this: A = _____
   b The two most populous countries in Asia are also the two most populous countries in the world. Name them.
   c Which is the *most* populous country in the world?
   d China is around half the size of Russia in area. But it has nearly ten times more people than Russia.
      i Which of the two has a lower *population density*?
      ii Explain your answer to **i** in terms of climate.

7 a Define *biome*.
   b The plants and animals in a biome depend on the *teamicl*. Unjumble the word in italics.

A

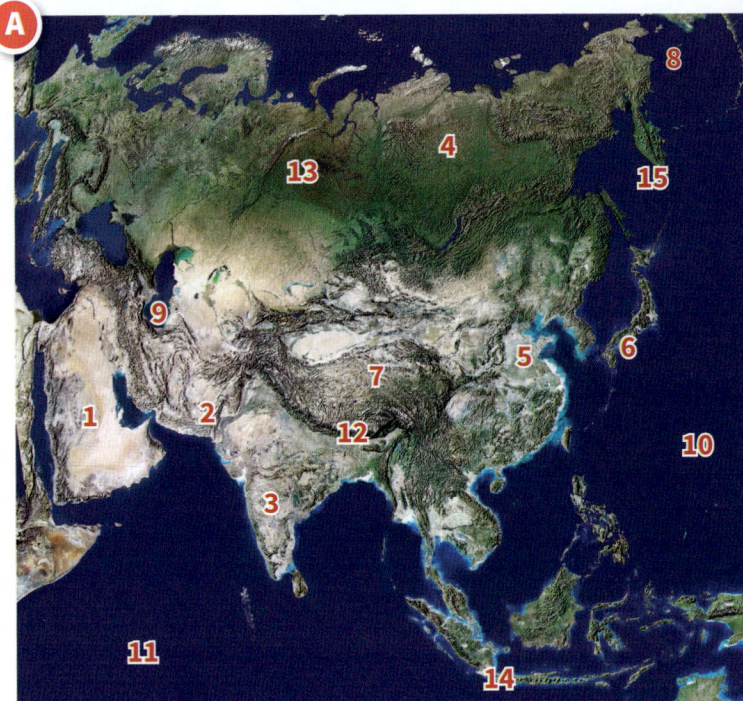

B

   c Here are four of Asia's biomes:
      taiga   steppe   tundra   cold desert
      Which of the four:
      i is coldest overall?   ii has most trees?
   d In which biome in **c** are you most likely to find:
      i horses?   ii polar bears?
   e Name the biome found around these letters on **A**:
      i 4   ii 1   iii 12   iv 14
   f Photo **B** shows a herder with his animals. In which country was this photo taken?
      i India   ii Russia   iii Yemen   iv Singapore
      Explain your choice.

8 *Asia is the world's most important continent.*
   To what extent do you agree with this statement? Give reasons to support your answer. Write at least half a page.

8 China

8.1 China: an overview

This unit introduces China – the world's most populous country, and one of the most important.

A big important country

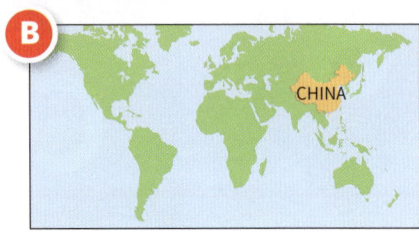

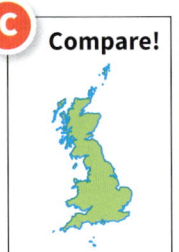

Compare!

▲ China's flag. Red is for the revolution which led to modern China. The big star represents the Communist Party, and the small stars the people.

China is a big, fascinating, important country.
- Its full name is the **People's Republic of China** (PRC). Its capital is Beijing.
- It has the world's largest population: over 1.4 billion people. Almost one in every five people in the world is Chinese!
- It's the fourth biggest country by land area. (After Russia, Canada, USA.)
- 500 years ago, China was a wealthy empire. Then it fell into poverty – but now it's catching up fast.
- Today it is the world's top exporter of goods. You are sure to own things made in China. Trainers, mobile, headphones, computer …

Hong Kong, Macau, and Taiwan

Find Hong Kong and Macau on the map. (Look along the coast.)
- Hong Kong was a British colony for 145 years, from 1842 to 1997. When it was handed back, China agreed to give it special status until 2047. So Hong Kong runs its own legal system, and economy.
- Likewise, Macau was a Portuguese colony for 442 years. It was handed back to China in 1999, and it too has special status.

Now find the island of Taiwan. It runs itself. But the People's Republic of China claims it, and wants unification. Many in Taiwan rebuff this claim.

▲ For over 2000 years, China was ruled by Emperors. This is the last Emperor, who was forced to step down in 1912, at age 6.

▲ China, 1967: heading for work on a communal farm. China was mainly agricultural then, with much poverty.

▲ Since 1978, China has undergone rapid industrialisation. Now it's the world's top manufacturer and exporter.

▲ China aims to lead the world in artificial intelligence and robotics. Here, a police robot patrols the streets of Beijing.

What's it like to live there?

If you are Chinese, living in China …

- you are one of over 1.4 billion people, as you saw earlier. (1 400 000 000.)
- you are likely to be from the Han ethnic group. Nearly 92% of the population is Han Chinese. (There are 55 minority ethnic groups.)
- you may be an only child. In 1979, the government set out a policy of one child per family. (Since 2016, parents can have two children – but many are happy with just one.)
- you may live in a town or city already – or will do later. 6 out of 10 Chinese people live in urban areas, and the number is rising fast.
- you study in Standard Chinese (a form of Mandarin) even if you don't speak it at home. And you may know more than 1500 Chinese characters!
- you are likely to have a higher standard of living than your parents did at your age. That's because China is growing wealthier every year.

▲ You have to work hard in school in China. Lots of homework, tests, and pressure! Many parents pay for extra tuition at weekends.

Your turn

1. Look at maps **A** and **B**. Page 141 may help too.
   a. Name the continent that China belongs to.
   b. i How many countries does China share a border with?
      ii Name the country it shares the longest border with.
      iii Name the largest country it shares a border with.
      iv Name four countries that border western China.

2. Map **C** shows Great Britain, for comparison. About how many Great Britains would fit into China?
   a about 5   b about 15   c about 45

3. About how wide is China at its widest, from west to east? Check the scale on map **A**, then choose.
   a about 4000 km   b about 7000 km   c about 2000 km

4. What does this represent, on China's flag?
   a the large star   b the red colour

5. Hong Kong consists of a group of islands plus a small part of mainland China. It has special links with the UK. Explain why.

6. Back to China's borders! Suggest one way that having borders with lots of other countries might:
   a benefit a country   b be a drawback for a country

7. a Which is the main ethnic group in China?
   b Why does China have so many children without siblings?

8. A hundred years ago China was poor. Now it's growing wealthier every year. Why? Suggest reasons.

8.2 A little history

 From empire to a modern republic, China has undergone enormous changes in the last 120 years. Find out more here.

> **Why ...** ... is china named after China?

From empire to a one-party state

For much of its recorded history, China had been ruled by Emperors. The first was Qin Shi Huangdi, above, who became Emperor in 221 BCE.

The title 'Emperor' was passed from father to son. But rebellions and invasions led to changes in China's dynasties – and borders – over time.

China grew wealthy and powerful. Science, philosophy and the arts were held in high esteem. There were many important inventions.

European traders bought silk, tea, porcelain and other Chinese goods. But China didn't want European goods. It wanted payment in silver.

So British traders began to sell opium into China. This shameful trade led to the first Opium War between Britain and China (1840 – 1842). China lost.

Other countries then forced China to sign trading treaties. Later, Britain and France fought the Second Opium War with China (1856 – 1860). China lost.

The defeats and hardship led to revolts within China. In 1912, the young Emperor was forced out. The Nationalist Party was set up.

But the Nationalists were unable to unite China. A second group – the Communists – struggled with them, and eventually took over.

In 1949, the Nationalist leaders fled to Taiwan, an island. The Communist Party took control of mainland China – and is still in control today.

Today, Taiwan calls itself the **Republic of China (Taiwan)**. But as you saw on page 120, the People's Republic of China claims it. This is not yet resolved.

China

The People's Republic of China

When the Communist Party took over, it aimed to make China a strong and self-sufficient modern industrial country. The leader was Mao Zedong. He named China the **People's Republic of China**.

Over time, the state took control of everything: land, factories, other businesses. People were told what work to do. The focus was on industry. Farmers were forced to work very hard to provide food for everyone. Some food was exported to earn money, even when yields were low.

Overall, it was not a success. In the period 1958 – 1961, the Chinese did not have enough food to feed themselves. Over 20 million people died of famine.

China today

When Mao Zedong died in 1976, China was still poor. It had not yet achieved its aim to become a modern industrial country.

The next leader made big changes. In 1978, Deng Xiaoping announced that China would open its doors to the world. **Special economic zones** were set up along the coast, to attract foreign companies. Chinese companies could team up with foreign ones. People could work to make money.

As a result, China developed rapidly through industry. The standard of living rose fast. You'll find out more in the rest of this chapter.

Still a one-party state

China is still run by the Communist Party. It's a one-party state. You can't vote for another party to take over.

The leaders plan China's development in detail. They have enormous power, so they can force change through fast, in this big country.

But China is no longer a truly communist country. It mixes communism with **capitalism**, where people are free to make money.

And China is now the country with most billionaires!

The key ideas of communism
- People who own factories and land use workers to make themselves rich.
- So their workers are like slaves.
- But we are all equal.
- So nobody should own property. The state should own everything.
- The state can then plan what to grow and make, to meet people's needs.
- The people can work to produce these things.
- In return they will get all they need, for free.

▼ A National Congress of the Communist Party is held every 5 years, and major decisions are announced. On the wall is the party's emblem: a hammer and sickle.

Your turn

1. For over 2000 years, China was an empire, ruled by an Emperor. How is China governed today?
2. Outline the causes of the First Opium War.
3. a What and where is Taiwan?
   b Explain why Taiwan's relationship with China is uneasy.
4. Read the panel about communism, above.
   a Identify any ideas you agree with.
   b Identify any ideas that you'd find difficult to accept.
   Each time, give your reasons.
5. a Define the term *special economic zone*. (Glossary?)
   b Suggest one reason why China's first special economic zones were set up along the coast.
6. a Is the UK a one-party state? Give evidence!
   b Try to suggest one advantage of being:
      i a one-party state ii a democracy (Glossary?)
7. Today China is not a truly communist country. Explain this.
8. Now start a spider map to summarise what you have learned so far about China. Use a double page so that you can add much more later. Use clear headings. Keep notes short!

8.3 Mainland China's physical geography

Find out here about mainland China's key physical features, and its climate.

Did you know?
- 'Mainland China' includes the tropical island of Hainan ...
- ... where Z is on map A.

It's big!

China is the fourth largest country in the world by land area, after Russia, Canada and the USA. It's very close in size to the USA. And it's about 40 times the size of the UK!

A Physical features of mainland China

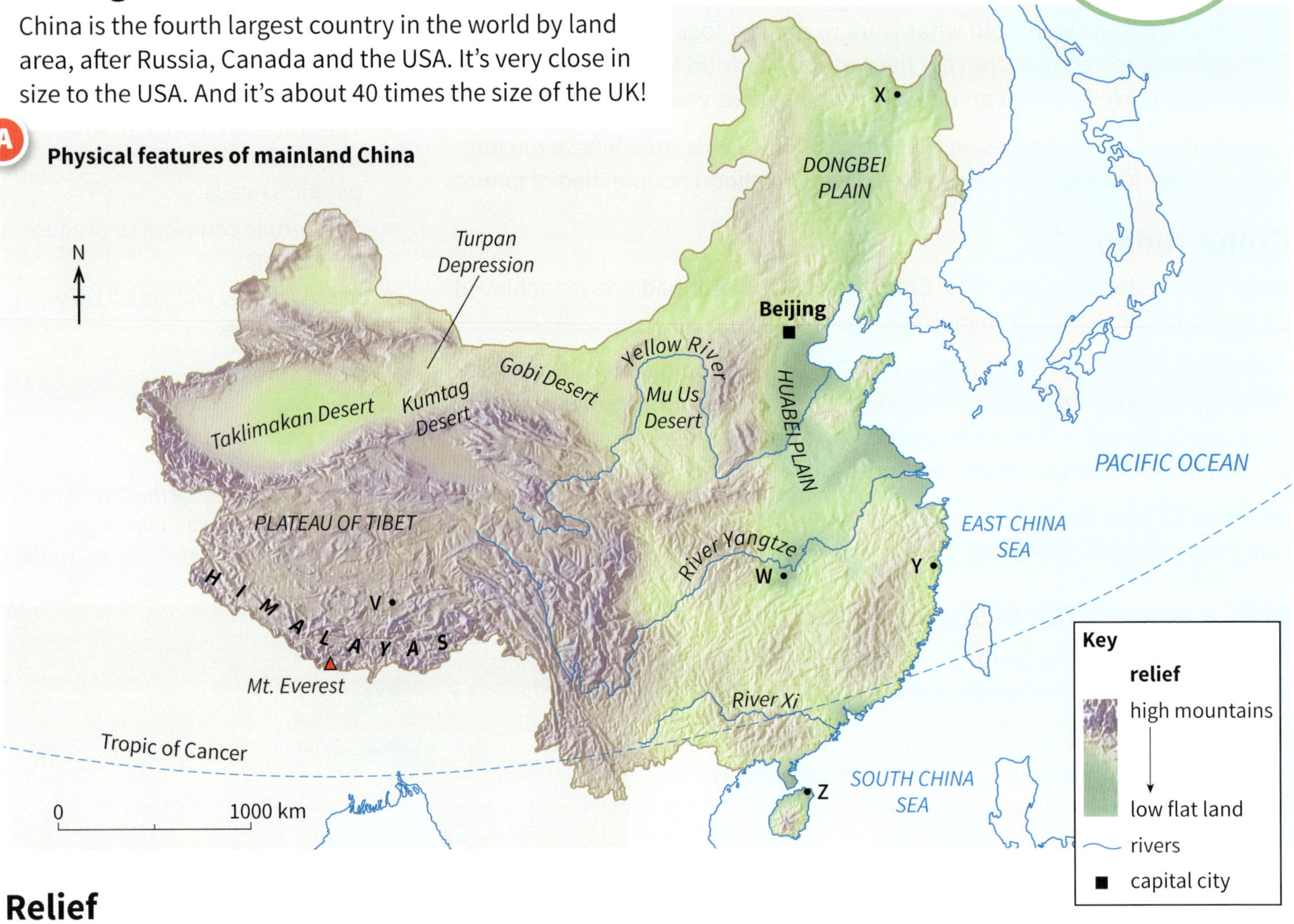

Relief

As you go west in China, the land steps upwards. Look at map **A**.

- The lowest areas are along the coast. The Huabei Plain is the largest area of flat land. It is very fertile. (Less than 13% of China's land can be farmed.)
- The next step up has mountain ranges, and large deserts. About 27% of China is desert. Some of the deserts are sandy, and some stony.
- The top step has the vast Plateau of Tibet. This is about 4000 m (or 4 km) above sea level, on average. The Himalayas form its southern border. Mount Everest lies on China's border with Nepal.

▲ A container ship on the Yangtze, China's longest river (6300 km).

◀ The world's highest railway on the world's highest plateau – the Plateau of Tibet.

China

China's main rivers

China has thousands of rivers. Map **A** shows just three, without their tributaries.

- The River Yangtze is China's longest river, and the third-longest in the world. It is also one of the busiest for river traffic.
- The Yellow River is China's second longest, and the sixth-longest in the world. It is named after the yellow silt it carries.
- The River Xi is very important for transport too. Along with several other rivers, it flows to the Pearl River Delta, one of China's top industrial areas.

Flooding has always been a problem on these rivers. Floods on the Yangtze are said to have killed over 2 million people in 1931. Dams on the Yangtze and Yellow River have helped to reduce flooding.

China's climates

China is vast, with complex **relief**. So it has a range of climates. Look at **C**.

- The north is sub-arctic. The far south is in the tropics. It is very cold on the Plateau of Tibet, since this is so high.
- Large land masses heat up fast in summer, and cool fast in winter. So inland, there's a big temperature difference between summer and winter.
- The sea keeps the coast cooler in summer, and warmer in winter.
- Inland, in summer, warm air rises. This draws moist **monsoon winds** into China from the oceans, bringing monsoon rain. Look at **B**.
- China's deserts are **cold deserts**. They get warm or hot in summer, but they are very cold in winter.

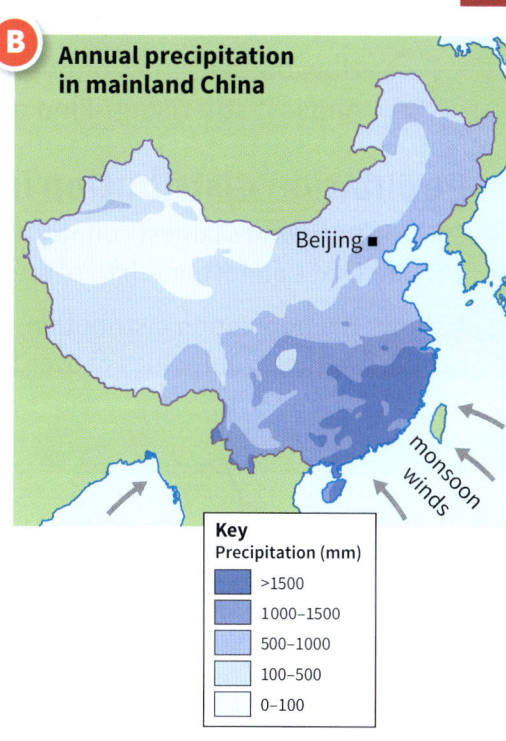

B Annual precipitation in mainland China

Key
Precipitation (mm)
- >1500
- 1000–1500
- 500–1000
- 100–500
- 0–100

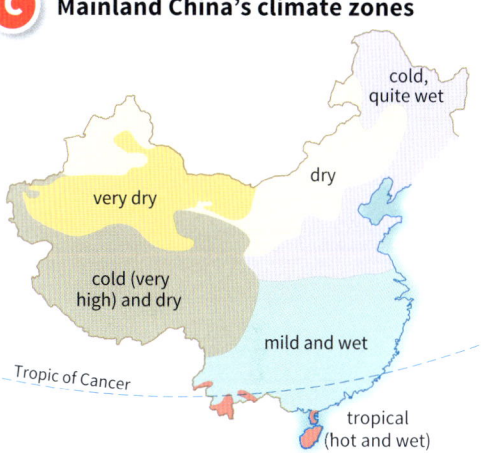

C Mainland China's climate zones

◀ In the Gobi desert in winter. Those are Bactrian camels (two humps).

Your turn

1. Name the seas off China, and the ocean they are part of.
2. Name China's longest river, and state its length.
3. a Define the term *relief*.
   b Describe the pattern of relief in China. In at least ten lines!
4. Look at places V – Z marked on map **A**. Explain why:
   a it is always cold at V
   b it is much warmer at W than at X
   c the temperature change from summer to winter is less at Y than at W
   d there are palm-fringed beaches and a warm sea at Z
5. a Using **C** to help you, describe the climate in each of V – Z.
   b Which place would you prefer to live in? Explain why.
6. a In China, most rain falls between May and October, in the *sonomon* season. Unjumble the word in italics.
   b What causes the rain during this season?
7. a Define the term *desert*. (Glossary?)
   b Explain why China's deserts are cold (unlike the Sahara).
   c The Taklimakan Desert gets less than 10 mm of rain a year. Use map **A** and page 141 to help you explain why.
8. Add information from this unit to your spider map for China.

8.4 Where is everyone?

How big is China's population? How is it spread aound the country? And is it still rising? Find out here.

Population distribution in mainland China

China is the world's largest country by population: over 1.4 billion people. That's almost one-fifth of the world's people.

A shows the population distribution. The deeper the shade, the more people live there. Only cities of over 1 million people are shown.

Did you know?
- In Chinese names, surnames are written first.
- A married woman keeps her own surname.

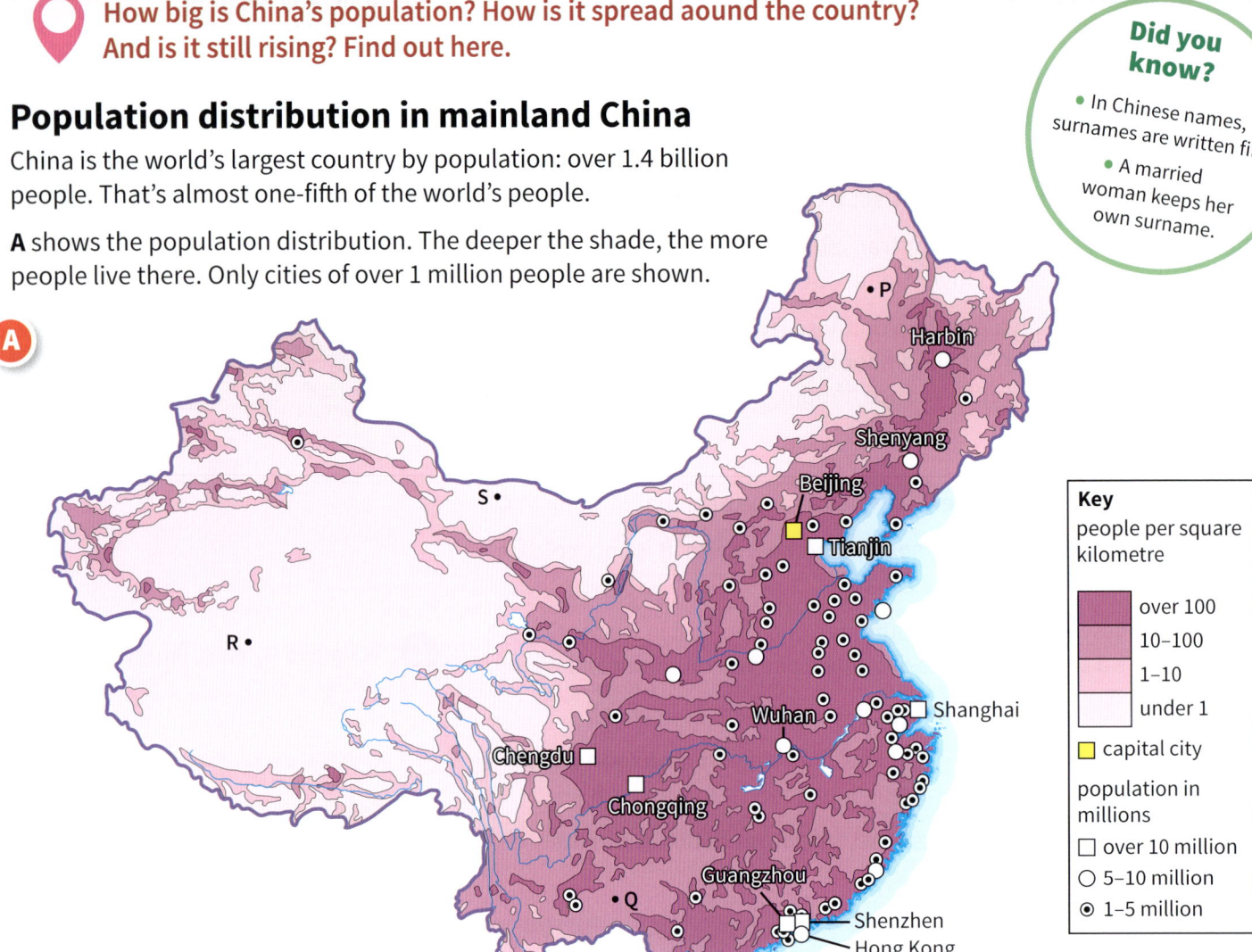

A

Key — people per square kilometre
- over 100
- 10–100
- 1–10
- under 1
- capital city
- population in millions
  - over 10 million
  - 5–10 million
  - 1–5 million

- Most of the population lives in the eastern half of China. Large parts of the country are nearly empty.
- This uneven distribution is partly due to China's physical geography. For example most of the fertile land is in the eastern half.
- The government had a role too. It encouraged factories to set up in towns and cities along and near the coast, to make exporting easier.
- Since 1978, hundreds of millions of people have migrated from rural areas, to find work in these factories. It's the biggest migration in history!

China's big cities

China has more than 100 cities of over 1 million! (The UK has 2.) And 10 megacities, with populations of over 10 million. (The UK has 0.) In fact it may have even more megacities, if all migrants were counted.

It's predicted that 1 billion Chinese people will be living in cities by 2030.

B **Population of the cities named on A**

City	Pop (millions)
Beijing	21.5
Chengdu	11.4
Chongqing	18.4
Guangzhou	11.5
Harbin	5.2
Hong Kong	7.5
Shanghai	24.2
Shenyang	8.1
Shenzhen	12.5
Tianjin	15.6
Wuhan	8.9

▲ Shanghai, China's biggest city.

▲ A rural village in southern China, set among rice terraces.

Is the population rising?

China's population is rising – but slowly. Look at **C**.

It is rising only slowly because the government has controlled its rise, to help reduce poverty.

In 1979 the government set out a **one-child policy**. A family could have only one child. There were fines and other punishments for people with more.

(There were exceptions. For example in a rural area, if your child was a girl, you could have a second child.)

It worked. The population rise slowed. But there were consequences:

- Too few girls were being born, because many parents chose to have boys.
- It became clear that there'd be too few workers by 2030, to support all the older people. (China now has an **ageing population**.)

So on 1 January 2016, the government changed the policy to allow *two* children per family.

But the population has not risen as much as expected since then. Many families are happy with just one child, or cannot afford more. The government may try to encourage more births in the future.

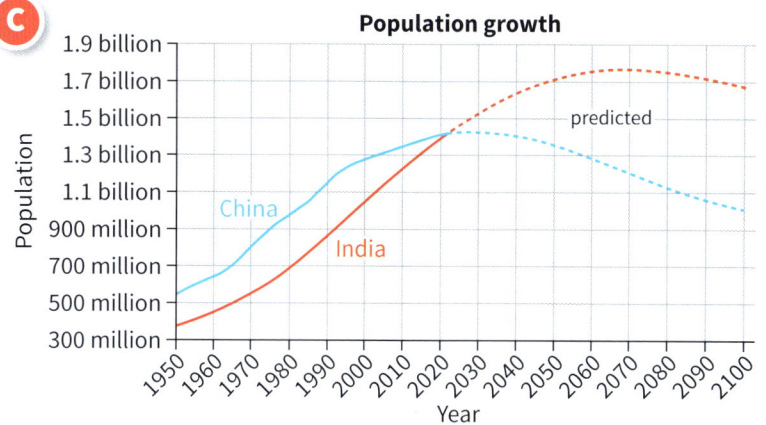

▲ A mural promoting the one-child policy.

Your turn

1. Look at map **A**. The population density around P is 1 – 10 people per square km. Think about it. That's low! What is the population density around: **a** Q? **b** R?

2. Using the information in Units 8.3 and 8.4, explain these:
   a. the low population density around: **i** R **ii** S
   b. the high population density in the triangle formed by Beijing, Shanghai, and Wuhan
   c. the high population density along the coast

3. a. Define the term *megacity*.
   b. Name China's biggest megacity.
   c. Name three megacities along the River Yangtze.

4. Graph **C** shows the population of China – and India too, for comparison. It includes predictions up to 2100.
   a. China's population has been rising more slowly than India's, since around 1980. Give a reason, from the text.
   b. How could a one-child policy help a country to reduce poverty? Explain.
   c. Around when is the population predicted to peak:
      i in China? **ii** in India?
   d. Population decline may cause problems for China. Why?

5. Do you think the one-child policy was fair? Give reasons. Think about its impact on parents *and* young people like you!

8.5 How Shenzhen became a megacity

 Many of China's cities have grown rapidly, thanks to industry. Here we take Shenzhen as example. (Pronounce it as *Shan-jan*.)

What if … … you learned Chinese?

Shenzhen, then and now …

Shenzhen, 1979. A city of around 0.3 million people (300 000 people). Connected by rail to Hong Kong, which was a British colony at that time.

Shenzhen, 2019. A megacity of about 12.5 million people. Built by migrants, and humming with innovation. All in the space of 40 years.

How did it happen?

In 1978, its leader decided that China must open up to the rest of the world, in order to climb out of poverty.

So, as a trial, four **special economic zones** were created along the south-east coast. The first was in Shenzhen.

These zones offered low tax rates, low land rents, and workers with very low wages.

So lots of companies set up in Shenzhen, to make things cheaply. Companies from Hong Kong, Japan, the USA, the UK, and other countries – and other parts of China.

Migrants flocked to Shenzhen, looking for work. The city grew and grew … and is still growing.

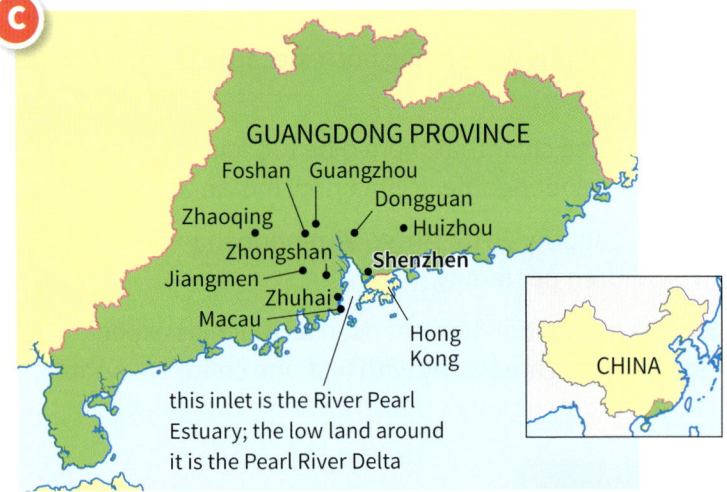

▲ Shenzhen is in southeast China, close to Hong Kong.

Changing industry

- At first, Shenzhen focused on making low-value goods, such as clothing, footwear and toys. Workers did not need much training.
- Then it became known for building or **assembling** higher-value goods, such as mobiles. The parts were often brought in from elsewhere.
- Now its focus is on high-tech research and innovation. It aims to lead the world in high-value areas like artificial intelligence, robotics, and gene research. It attracts highly-educated and highly-skilled workers.

▶ In a smartphone factory in Shenzhen.

▲ Behave! If you cross against the lights on this busy road in Shenzhen, your photo will pop up on that screen. If the system has data on you (which is likely) you'll get a text message, and a fine!

▲ All buses and taxis in Shenzhen are electric. And its industry is **light industry**. So the air is cleaner than in many other Chinese cities. This bus on a trial run goes one step further – no driver!

The migrant workers

Shenzhen has always depended on migrant workers, mostly from rural areas.

- Many have poor education, and get only low-skilled jobs in factories, canteens and so on.
- They don't earn much – although wages in Shenzhen are higher than in most parts of China.
- They have **hukou** (registration) in their home areas. So they are treated as temporary workers in the city. They have limited access to doctors, schools, and other services. That makes life harder.
- Some cities now offer city hukou to all migrants. But in Shenzhen, it depends on your education and work experience. Low-skilled migrants lose out!

▲ Migrant factory workers usually live in company dorms, because they can't afford to rent rooms outside the factory. Most save up to send money home.

Where next for Shenzhen?

Look at **C**. The eleven named cities cluster around the Pearl River Delta. Together they form the **Greater Bay Area**. In 2019, China launched a plan to improve links between these cities, to help the whole area develop.

The cities will work together, instead of competing. Shenzhen will act as the high-tech hub. Together, the cities will form a huge **megalopolis**.

Did you know?
- When you apply for a new mobile, or internet access, in China, your face is scanned.
- The image will be held in a data bank.

Your turn

1. Where in China is Shenzhen? Give as much detail as you can.
2. Study photos **A** and **B**. Then write three sentences comparing Shenzen in 1979 and 2019.
3. One reason foreign companies came to Shenzhen is this: workers' wages were far lower than in their own countries. Explain why low wages would appeal to companies.
4. Explain how the special economic zone in Shenzhen helped to lift Chinese people out of poverty. You could answer with a flowchart, or spider map, or drawings, or bullet points.
5. Shenzhen has *light industry*. Explain this term. (Glossary?)
6. Migrant factory workers usually live in dormitories, like in **G**.
   a. Give three ways in which dorm life might be difficult.
   b. They are willing to put up with dorms. Give two reasons.
7. a. What is *hukou*? (Glossary?)
   b. Having rural hukou makes life harder in Shenzhen. Why?
8. a. What is a *megalopolis*? (Glossary?)
   b. Name four cities in the Greater Bay Area megalopolis.
9. Give one example from this unit to show that Shenzhen is now a city of innovation. (Check the photos!)

8.6 Life in rural China

 Over 500 million people live in China's rural areas. What is life like for them? Find out here.

If you live in rural China …

China has been **urbanising** fast since 1979. Even so, it still has a large rural population: over 500 million people.

If you live in rural China …
- your family is likely to be in farming – although there is some rural industry.
- the farm is likely to be small, so you can't grow that much.
- your family will not own the land. It leases it. (In China, the land is collectively owned.)
- your family may be quite poor. People in rural areas earn much less than people in urban areas, on average.
- you probably won't have good services. Schools and hospitals are likely to be poor quality. There may be no public transport.

The farming challenge

China's vast population needs food. So agriculture is very important. And especially when less than 13% of China's land can be farmed.

On average, farms are small. (Before 1978 there were large communal farms, where people shared the work. After 1978 these were broken into small plots and leased to households – and people could sell their crops.)

The government is encouraging larger farms, so that people can grow more.

For example you can offer your land to a 'land bank', which pays you rent. Other farmers, or **agribusinesses**, then lease it from the land bank, so that they can grow more food, more efficiently, on larger farms.

China has big plans to improve roads, healthcare, and other services in rural areas.

Farming in China

Farmers grow rice, wheat, potatoes and other vegetables, apples, oranges and other fruit, peanuts and other seeds, tea, and cotton.

Farmers raise chickens, ducks, pigs, cattle, sheep, goats, horses and donkeys, and yaks and camels in some areas.

▲ Women planting rice in the paddy fields. Rice is China's main crop. It's grown in the areas with mild climates.

◄ Over 90% of farms in China are very small – smaller than a football pitch. So you can't use large machinery.

▼ China still has millions of people living in great poverty in rural areas. It aims to end poverty everywhere.

Off to the city ...

The Chinese government aims to improve life in rural areas.

But it also wants urbanisation to continue. Jobs in urban areas pay more than farming, and help lift people out of poverty.

Many young people leave family farms to go off to the city to earn a living. They send money home. These **remittances** are a big help to rural families.

Many migrants then send their children back home to be brought up by grandparents. That's because life in the city is expensive – and in China's largest cities, rural migrants do not have equal access to all services.

▲ Many rural villages in China have mainly elderly people and their grandchildren.

He Chan, a left-behind

I'm Chan. I'm ten. My mum and dad went to Chongqing city to work, so they left me behind in the village, with my gran. Children like me are called the left-behinds.

I love gran, but I miss mum and dad. When they come here they bring me presents, and clothes. They always come for the Chinese New Year. It takes them two days to get here!

I talk to them on the phone a lot. They always ask about school, and whether I have done my homework. I try to get a good school report, to please them.

When I am older I will go to the city. I'll take gran with me. We can all be together. I'd like that so much.

Which city?

As you saw above, low-skilled workers from rural areas get only limited access to services – such as healthcare and education – in China's largest cities.

But they are given equal access to all services in other cities. This is to encourage them to go these cities, to help balance the population distribution.

Did you know?
- China has around 200 million farms.
- The UK has around 0.2 million.

Did you know?
- The Chinese New Year festival lasts for 15 days.
- Its dates change across January and February, depending on the full moon.

Your turn

1. People complain about the poor education in rural schools in China. Give three reasons why good teachers may prefer to work in urban areas.
2. Describe the rural scene in photo **A**.
3. a A rice plant gives 10 g of rice. If you have 70 g of rice in your lunch, how many rice plants does that take?
   b Is rice farming easy work? Give your evidence.
4. a Define the term **agribusiness**. (Glossary?)
   b The Chinese government is now encouraging bigger farms, and agribusiness. Explain why.
5. Study the photos in this unit. What evidence can you see that there is poverty in rural areas?
6. China has 69 million left-behind children like Chan, according to a United Nations organisation.
   a Why are they called *left-behind*?
   b List the pros and cons of leaving children behind ...
      i from the parents' point of view
      ii from the children's point of view
7. How might rural life in China have changed by 2040? Think about the size and age of the rural population, services, jobs available, and so on. Explain each change you list.

8.7 What about the environment?

Here you'll learn about the state of the environment in China, and how it's being improved.

Develop first, clean up later

In Britain, industry took off with the Industrial Revolution. And so did pollution. Factory chimneys belched soot and dust and harmful gases from the coal-fired furnaces. Factory waste was dumped in rivers.

And then, as Britain grew more developed, we began to clean up.

China is following a similar pattern.

So what about the environment in China?

China has developed rapidly, and the environment has suffered badly.

- China is the world's top emitter of **carbon dioxide**, linked to climate change. This gas is produced by burning fossil fuels – coal, oil and gas. (But China is not top when you calculate emissions *per person* …)
- The air, soil and water have been polluted over decades by fumes, soot and other particles, and waste chemicals from factories.
- In rural areas, rivers and wells are polluted by fertiliser and pesticides.
- **Desertification** is also a problem in northern China. The Gobi Desert has spread into farmland. This is because trees have been chopped down, and land overgrazed, and there's less rain, thanks to climate change.

Coal, a major culprit

China's development is fuelled by coal, like the UK's was. By 2018, nearly 60% of China's energy still came from coal. It is burned in power stations and factory furnaces. It is the dirtiest fossil fuel, causing the most pollution.

▼ Air pollution in Beijing. It's a big problem in many of China's cities. In the background is part of the Forbidden Palace, which was once home to emperors.

▲ A coal-burning power station. The tall thin chimney gives out harmful gases and soot, which damage lungs, plus carbon dioxide. The fat chimneys give out steam.

▲ This is what desertification looks like. They are trying to restore the ruined farmland.

▲ Fighting back against air pollution! There are car-sharing schemes for electric cars in many Chinese cities.

Did you know?
- Industrialisation has helped to lift over 850 million people out of extreme poverty, in China.

So ... is China cleaning up?

China is doing a lot to tackle pollution.

- Many polluting factories have been shut down.
- China now gets more energy from dams, and wind and solar farms, than any other country. (Even so, these provide only a small share of its energy.)
- It has more electric cars in use than the rest of the world combined!
- But there's a big problem. China is burning more and more fossil fuels – for electricity, heating, transport and so on. Look at table **F** below. So its emissions of carbon dioxide keep rising.

▲ The Three Gorges dam on the Yangtze gives hydroelectricity – and helps to control floods. China has over 23 000 large dams (over 15 m high)! And many small ones.

China's sources of energy

Energy source	Amount of energy provided by this source (Mtoe) ...	
	in 2008	in 2018
coal	1609	1907
oil	385	641
gas	70	243
nuclear	15	67
renewables	151	416
Total energy used	2230	3274

▲ Fitting solar panels on a building in Wuhan city.

What about desertification?

China is tackling desertification by planting *billions* of trees and shrubs. The project is called **the Green Great Wall**. It began in 1978 and will continue until 2050.

It has been successful in some places – but not all. Many millions of trees have died from lack of water. Trees also soak up groundwater, leaving other places drier.

So China is looking at other solutions too.

People have been moved off desertified land to allow tree planting. China calls them **climate refugees**.

▲ Trees planted on desertified land, as part of the Green Great Wall.

Your turn

1. Look at the photos in this unit. What evidence can you see:
   a. that China suffers from air pollution?
   b. that China is taking steps to fight air pollution?

2. Look at table **F**. *Mtoe* stands for *millions of tonnes of oil equivalent*. It lets us compare energy from different sources.
   a. How did *total* energy use change between 2008 and 2018?
   b. Suggest one reason for this change. (Unit 8.5 may help.)
   c. Renewables include hydro, wind power and solar power. Explain why these are called *renewables*. (Glossary?)

3. Look again at table **F**.
   a. i Identify the main fuel used in China in both years.
      ii Compare the use of this fuel in 2008 and 2018.
      iii State one *local* and one *global* impact of this fuel.
   b. What evidence is there that China built more dams, solar farms and/or wind farms between 2008 and 2018?

4. Outline what China is doing to fight desertification.

5. China put industrialisation before protecting the environment. Was this the right choice? Explain.

133

8.8 What's the Belt and Road Initiative?

China has a big ambitious project which will affect much of the world. Found out more about it here.

What's the Belt and Road Initiative?

The **Belt and Road Initiative** or **BRI** is a project to create a network of links from mainland China across the world. Roads, railways, pipelines, ports …

China announced it in 2013. It hopes to complete it by 2049.

This map shows the links. The **Belt** is the land routes. The **Road** is the sea routes (which is a little confusing).

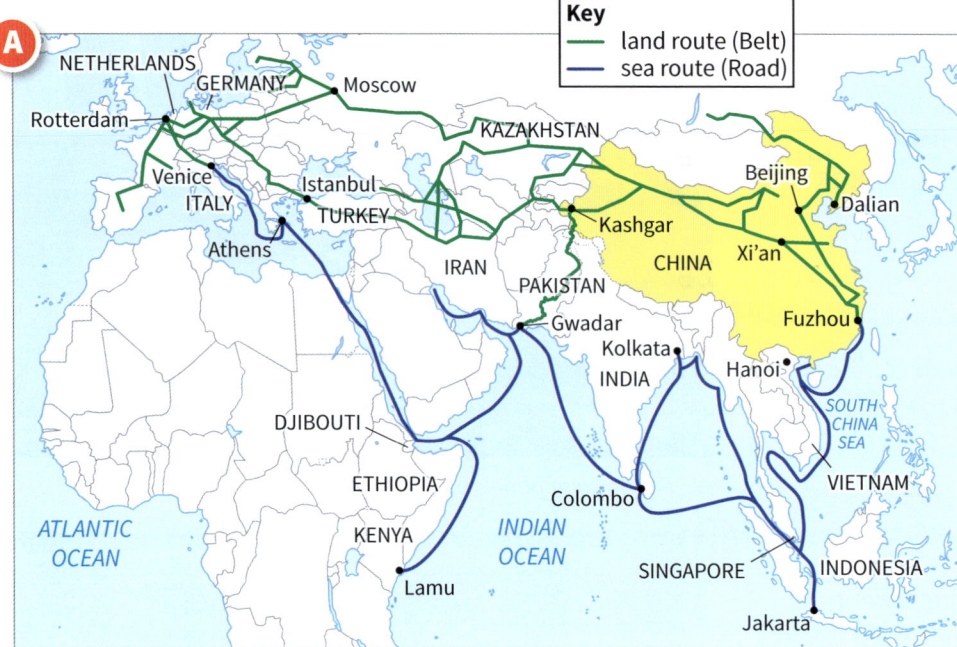

A

▲ These coffee beans harvested in Ethiopia will travel to China via Djibouti.

B

▲ The overland route from the port of Gwadar in Pakistan to Kashgar in China.

Why is China doing this?

- First, as the world's top manufacturer, China has earned a lot of money – and wants to invest it.
- The BRI will help China to export and import goods more easily, and create jobs for Chinese workers, and develop western China.
- But China says the project will benefit the whole world, not just China. It will speed up development in many countries.
- Some countries are wary. They think China aims to increase its power and influence around the world. China rejects this view.

Example: Pakistan in the BRI

In Pakistan, China is building a port and an airport at Gwadar. (Find it on **A**.) Plus a special economic zone for factories near the port, and a transport network to link Gwadar to Kashgar in western China. Look at **B**.

Why? Suppose China wants to import food and other things from Africa. It can ship them to Gwadar, then take them by land to Kashgar. That's a lot shorter than a sea journey from Africa to China's east coast. This route would also bypass any conflicts in the South China Sea.

▲ A truck on the highway leading from Pakistan to Kashgar in China.

This is how BRI works …

China talks to countries about BRI projects. They are usually big costly **infrastructure** projects – to build ports, railways and roads, pipelines for oil or gas, *and* power stations that burn fossil fuels.

▼

China offers to lend the countries money for the projects, at low interest rates, if they need it. Countries – and especially lower income countries – may be very happy to agree.

▼

Chinese companies arrive to carry out the projects, where needed.

▼

Everyone looks forward to the benefits – including China.

By 2019, over 125 countries had agreed to cooperate with the BRI plan.

The trouble is …

A big costly project may take a long time to benefit a country. But the country still has to pay back any loan to China.

So far, several countries have been unable to pay. So China has reduced or cancelled some debt. It has also given countries more time to pay – for example 30 years instead of 10.

A country may still have difficulty. For example in 2016, Sri Lanka had huge debts, which it could not repay. So it leased to China, for 99 years, a port which China had built for it, plus a large area of land.

The port is at Hambantota in **C**. Some say it was not needed, because there's a port at Colombo. India and other countries are worried about it. They fear it could be used for Chinese warships one day.

The future …

The BRI has already helped some countries. It is too soon to say how successful the whole project will be. It is likely to shake up world trade. But it may also mean a rise in the use of fossil fuels!

▲ *China built a commercial port in Djibouti as well as a base for its navy.*

▲ *Sri Lanka.*

▲ *Chinese workers building a new railway station near Hambantota in Sri Lanka.*

Your turn

1. Explain what the *Belt and Road Initiative* is.
2. Using map **A** to help you, name any 5 countries which are on:
   a. a BRI land route
   b. a BRI sea route
3. China will spend many trillions of dollars on the BRI. Where does China get all this money?
4. Give two ways in which the BRI benefits China.
5. The Chinese government wants to develop industry in the west of China. Explain how the BRI link from Gwadar to Kashgar may help. (See map **B**.)
6. China has built a port at Lamu in Kenya. Kenya exports tea.
   a. Describe two BRI routes that a cargo of tea could take from Lamu to Kashgar in western China. (One via Fuzhou!)
   b. Which of your two routes in **a**:
      i. is shorter?   ii. avoids conflict in the South China Sea?
7. Suggest one way that the special economic zone at Gwadar will benefit Pakistan.
8. Overall, is the BRI a good idea? Discuss! Think about its impact on China, other countries, *and* the environment.

8 China

How much have you learned about China? Let's see.

check ✓

1 Map **A** shows mainland China.
  a Dots 1 – 5 represent cities. Which dot represents:
    i Beijing? ii Kashgar?
  b i Which dot represents China's largest city?
    ii Name this city.
  c i Which dot represents a city that was at the heart of a small British colony until 1997?
    ii State the name of this city.
  d Look at the areas numbered 6 – 8. Which one is:
    i in a cold desert? ii the most highly industrialised?
    iii the most highly populated? Explain each choice.
  e i Define the term *plateau*.
    ii Name the plateau that includes the area labelled 8.
  f Name China's two longest rivers (not shown on the map).
  g Explain in three sentences why China gets monsoon rain.

2 Graph **B** shows the *average* wealth per person in China. It is expressed as GNI per person, in dollars.
  a Describe the trend shown in the graph.
  b Roughly what was the GNI person: i in 2000? ii in 2010?
  c Suggest one reason why GNI per person is growing.
  d The Belt and Road Initiative may increase the GNI per person even further. Explain why.

3 This question is about the photos on page 119.
  a Photo **C** shows Chongqing city. This photo was chosen to represent urbanisation. Define *urbanisation*.
  b Photo **D** shows a scene from rural China. (The tall shapes in the distance are *karst* landforms, made of limestone.) State two facts about life in rural China.
  c Give one pull factor that attracts farmers like those in **D** to a city like Chongqing.
  d China's national animal feeds on bamboo. Name it. (Photo?)
  e One photo shows a key reason for China's growing power in the world. Identify the photo, and explain your choice.
  f The people in **F** are doing Tai Chi, an ancient Chinese exercise program. It is popular in the UK too. Name one other aspect of Chinese life and culture that is popular in the UK.

4 Table **C** below shows trade with China for two countries in 2018.
  a Name one thing you own, that was made in China.
  b How much did the UK pay for imports from China in 2018?
  c In their trade with China that year, who earned more?
    i China, or the UK? ii China, or the USA?
  d Being the world's top exporter is making China wealthier. Explain why, using **C** to help you.

C Value of trade with China in 2018

Country	Exports to China	Imports from China
UK	£22.6 billion	£44.7 billion
USA	$120 billion	$540 billion

A Mainland China

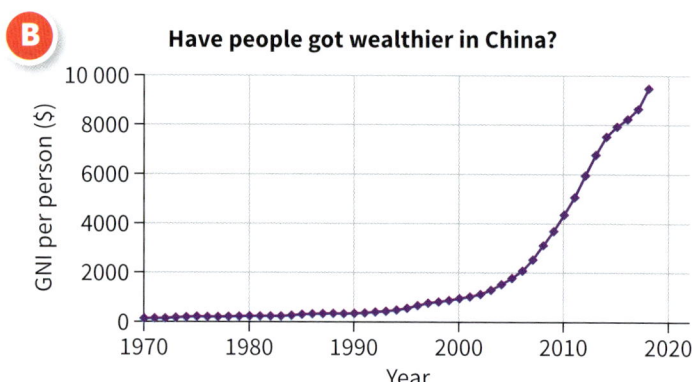

B Have people got wealthier in China?

5 China consumes vast amounts of energy, for factories, transport, and so on. **D** shows the sources of this energy in 2018.

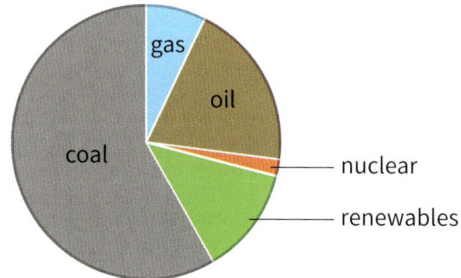

D China's energy sources in 2018

  a Name three renewables China uses. (Glossary?)
  b This pie chart is a worry for the world. Explain why.

6 *China has a big impact on the rest of the world. Do you agree with this statement? Decide, and give evidence to support your answer. Write at least half a page.*

Command words: a summary

All through *geog.123* you'll meet questions with command words that tell you how to answer them. Later you'll meet them in exams too. So it's good to get used to them.

Here's a summary of the command words to help you, in alphabetical order. The command words and their definitions are in red. (And note that we cover them in more detail in Chapter 1 of *geog.1*.)

Assess
Weigh everything up and make a judgement.
For example assess the impact of an earthquake on a city. You must always include the evidence you based your judgement on.

Calculate
Do some maths, to get the answer!
Always put the unit in your answer. For example *5 km* or *11 people* or *15 days*. (Not just *5* or *11* or *15*.)

Compare
Identify what is the same, and different, about two things.
For example say which one is bigger. Always mention *both* things in your answer.

Copy and complete
Copy this, filling in all the blanks.
Fill in using the words and terms that make sense!

Define
Write down the meaning.
Keep your answer clear and simple.

Describe
Write a description.
For example describe what you see, or the steps in a process. You do not need to give reasons for anything.

Discuss
Look at something from different angles, and give key points about it.
For example you could give its good and bad points, or its benefits and drawbacks.

Draw
Like it says – draw!
For example draw a diagram, or sketch map, or bar chart, or line graph. Use a ruler for straight lines. Be accurate with bar charts and graphs. And try to be quick!

Evaluate
Judge how successful or worthwhile something is.
You should say what has been good and bad about it, *and* give your final opinion.

Examine
Look at each part and say how it contributes.
For example examine how different processes work together to form an oxbow lake.

Explain
Make something clear and easy to understand.
For example explain how a meander forms.

Give
Come up with an answer, from what you've learned.
Keep it clear and simple.

Identify
Pick out the thing, and give its name.
For example identify a landform on a map.

Justify
Give reasons to support the choice or decision you made.
For example give reasons why you agree with a statement.

Label
Add labels!
For example label a diagram. The aim is to make it clear and easy to understand. So keep labels short and simple!

Name
Write the name of the thing you are asked about.
Easy! You do not need to write a full sentence.

Outline
Set out the main points.
Stick to the main points. You don't need to give details.

State
Give the answer in clear terms.
State is often used in place of *Give* or *Identify* or even *Calculate* or *Count*. Be sure to answer clearly.

Suggest
Come up with a possible reason or plan.
Use your common sense!

To what extent .... ?
How much does it contribute, or how important / true is it?
Make a judgement, and give the evidence you based it on.

Ordnance Survey symbols

ROADS AND PATHS 1: 25 000

- M 1 or A 6(M) — Motorway
- A 35 — Dual carriageway
- A 30 — Main road
- B 3074 — Secondary road
- Narrow road with passing places
- Road under construction
- Road generally more than 4 m wide
- Road generally less than 4 m wide
- Other road, drive or track, fenced and unfenced
- Gradient: steeper than 1 in 5; 1 in 7 to 1 in 5
- Ferry; Ferry P – passenger only
- Path

PUBLIC RIGHTS OF WAY

1:25 000	1:50 000	
		Footpath
		Bridleway
+++++	− − − −	Byway open to all traffic
+-+-+-+	-+-+-+-	Restricted bridleway

RAILWAYS 1: 25 000

- Multiple track
- Single track
- Narrow gauge/Light rapid transit system
- Road over; road under; level crossing
- Cutting; tunnel; embankment
- Station, open to passengers; siding

BOUNDARIES 1: 50 000

- National
- District
- County, Unitary Authority, Metropolitan District or London Borough
- National Park

HEIGHTS/ROCK FEATURES 1: 50 000

- Contour lines
- •144 Spot height to the nearest metre above sea level
- outcrop, cliff, scree

ABBREVIATIONS 1: 25 000 and 1: 50 000

PO / P	Post office	PC	Public convenience (rural areas)
PH	Public house	TH	Town Hall, Guildhall or equivalent
MS	Milestone	Sch	School
MP	Milepost	Coll	College
CH	Clubhouse	Mus	Museum
CG	Cattlegrid	Cemy	Cemetery
Fm	Farm	Hosp	Hospital

ANTIQUITIES 1: 25 000 and 1: 50 000

- VILLA Roman
- Castle Non-Roman
- Battlefield (with date)
- Visible earthwork

LAND FEATURES 1: 25 000 and/or 1: 50 000

- ruin — Buildings
- Public building
- Bus or coach station
- Place of Worship (current or former) with tower / with spire, minaret or dome / without such additions
- Chimney or tower
- Glass structure
- Heliport
- Triangulation pillar
- Mast
- Wind pump / wind turbine
- Windmill
- Graticule intersection
- Cutting, embankment
- Quarry
- Spoil heap, refuse tip or dump
- Coniferous wood
- Non-coniferous wood
- Mixed wood
- Orchard
- Park or ornamental ground
- Forestry Commission access land
- National Trust – always open
- National Trust, limited access, observe local signs
- National Trust for Scotland

WATER FEATURES 1: 25 000 and/or 1: 50 000

- Marsh or salting, Towpath, Lock, Slopes, Cliff, High water mark
- Aqueduct, Canal, Ford, Flat rock, Lighthouse (in use), Low water mark
- Lake, Weir, Normal tidal limit, Sand, Dunes, Lighthouse (disused), Beacon, Shingle
- Footbridge, Bridge, Mud
- Canal (dry)

TOURIST INFORMATION 1: 25 000 and/or 1: 50 000

- P Parking
- V Visitor centre
- i Information centre
- Recreation/leisure/sports centre
- Telephone
- Camp site / Caravan site
- Golf course or links
- Viewpoint
- PC Public convenience (toilet)
- Picnic site
- Pub/s
- Cathedral/Abbey
- Museum
- Castle/fort
- Building of historic interest
- English Heritage
- Garden
- Nature reserve
- Water activities
- Fishing
- Other tourist feature

© Crown copyright

Map of the world

— international boundary
• capital city

abbreviations
BELG.	BELGIUM
B-H.	BOSNIA-HERZEGOVINA
C.	CROATIA
CENT. AF. REP.	CENTRAL AFRICAN REPUBLIC
CZ.	CZECH REPUBLIC
F.	FYROM (Former Yugoslav Republic of Macedonia)
K.	KOSOVO
LITH.	LITHUANIA
MT.	MONTENEGRO
LUX.	LUXEMBOURG
NETH.	NETHERLANDS
S.	SLOVENIA
SE.	SERBIA
SL.	SLOVAKIA
SWITZ.	SWITZERLAND
U.A.E.	UNITED ARAB EMIRATES
U.S.A.	UNITED STATES OF AMERICA

Equatorial Scale 1: 95 000 000

Did you know?
- Earth is 4600 million years old.
- It weighs 6000 million million million tonnes.

The continents and oceans

Amazing – but true!
- Nearly 70% of Earth is covered by saltwater.
- Nearly 1/3 is covered by the Pacific Ocean.
- 10% of the land is by glaciers.
- 20% of the land is covered by deserts.

World champions
- Largest continent – Asia
- Longest river – The Nile, Africa
- Highest mountain on land – Everest, Nepal
- Highest mountain in the ocean – Mauna, Hawai
- Largest desert – Sahara, North Africa
- Largest ocean – Pacific

Did you know?
The world has:
- over 190 countries
- nearly 8 billion people
- over 6000 different languages

141

Glossary

A

abrasion – scraping away material

agribusiness – a business based on farming

air mass – a huge block of air moving over Earth; it can be warm or cold, damp or dry, depending on where it came from

air pressure – the force pressing down on us due to the weight of the atmosphere

altitude – height of a place above sea level

arch – the curved structure left when the sea erodes the inside of a cave away

atmosphere – the layer of gas around Earth

B

bay – a smooth curve of coast between two headlands

beach – an area of sand or small stones, deposited by waves

beach replenishment – adding sand to a beach to replace sand the waves carried away

biome – a very large area with a similar climate, plants, and animals, throughout

C

capital city – where the government is based

carbon neutral – adding no extra carbon dioxide to the atmosphere; to achieve this we could stop using fossil fuels, or remove all the carbon dioxide they produce (for example by planting forests)

cave – a large hollow in rock, for example in the side of a cliff

cliff – a big mass of rock with a steep face; there are many cliffs along the UK's coast

climate – what the weather in a place is *usually* like, over the year; they take measurements over long periods and calculate the average

climate change – all aspects of climate are changing because Earth is getting warmer

coast – where the land meets the sea

coastal defences – barriers to protect the coast from erosion or flooding

condense – to change from gas to liquid

coniferous – describes trees which bear cones (such as fir trees)

continent – one of Earth's great land masses; there are seven continents

convectional rainfall – the Sun heats the ground, convection currents of warm air rise, the water vapour condenses, and rain falls

country – humans have divided continents into political units called countries

D

dam – a structure built across a river to control water flow; it usually has turbines which the water spins, generating electricity

data – information collected for a purpose; for example measurements of a river's depth

deciduous – describes trees which lose their leaves in winter

decline – to fall gradually into a poor state

de-industrialisation – loss of industry

democracy – where people choose their government, by voting

densely populated – lots of people live there

deposit – to drop material; waves deposit sand and small stones to form beaches

depression – a weather system made up of a warm front chased by a cold one; it brings wet windy weather

desert – gets very little rain; it can be a hot or cold desert, and sandy or rocky

desertification – where farmland becomes like a desert, through overuse or drought

development – a process of change that goes on in a country, with the aim of improving people's lives

drought – there is less rain than usual, so there is not enough water for our needs

E

economic – about money and business

economy – all the business activity going on in a country (in producing, transporting, selling, and buying things)

emissions – waste gases that go into the air, for example from car exhausts

endangered – when so few of a species are left that it's in danger of becoming extinct

environment – everything around you; air, soil, water, animals, and plants form the natural environment

Equator – an imaginary line around the middle of Earth (at 0° latitude)

erosion – the wearing away of rock, stones and soil by rivers, waves, wind, or glaciers

exports – things sold to other countries

F

fertility rate – the average number of children per woman, in a population; it's an average, so it is not usually a whole number

fetch – the length of water the wind blows over, before it meets the coast

fieldwork – where you go out and collect data, to test a hypothesis or answer a question

flood defences – structures built to prevent flooding; for example an embankment

fossil fuels – coal, oil, natural gas

front – the leading edge of an air mass; a warm front means a warm air mass is arriving

frontal rainfall – rain caused by a warm front meeting a cold one

G

GIS (geographic information system) – lets you display layers of data on a map on a computer screen, to help you make decisions

glacier – a river of ice

global warming – the rise in average temperatures around the world

GNI (gross national income) – a country's total income for a year, given in dollars

GNI per person – a country's GNI divided by the population; it's a measure of how wealthy the people are, *on average*

GNI per person (PPP) – the GNI per person is adjusted to take into account that things cost more in some places than others

GPS (global positioning system) – tells you exactly where on Earth you are, using radio waves from satellites

gravity – the force of attraction that draws things towards Earth

greenhouse gases – they trap heat around Earth

groyne – barrier of wood or stone down a beach, to stop sand being washed away

H

headland – land that juts out into the sea

hemisphere – half of Earth; the Northern Hemisphere is the half above the Equator

hukou – registration; in China you have rural or urban hukou depending on your birthplace

hunter-gatherers – they lived by hunting animals and collecting fruit and seeds

hurricane – a violent spinning storm that starts in warm tropical waters; other names for it are *tropical cyclone* and *typhoon*

hydro – short for *hydroelectricity*

hydroelectricity – electricity generated when flowing water spins a turbine, at a dam

hypothesis – a statement that you can test, to see if it's true, by analysing data

I

ice age – when Earth's average temperature is low, and glaciers spread

imports – things bought from other countries

Industrial Revolution – the period (about 1760 – 1840 in Britain) when many new machines were invented, and factories built

industry – manufacturing, for example *the car industry*; the term is often used for services too, for example *the tourism industry*

inequality – when wealth and opportunity are not shared fairly among people

insulated – protected by material that stops heat escaping

Glossary

isobar – line on a weather map joining places at equal air pressure

L

landfill sites – big holes in the ground where household rubbish is buried; the UK has over 500 of them

latitude – how far a place is north or south of the Equator; it is measured in degrees

leeward – sheltered from the wind

life expectancy – how many years a new baby can expect to live for, on average

light industry – makes small things; *heavy industry* makes things like steel and ships

longitude – how far a place is east or west of the Prime Meridian; it is measured in degrees

longshore drift – how sand and other material is carried parallel to the shore, by the waves

M

mangroves – trees that grow in salty swamps along the coast

manufacturing – making things in factories

megalopolis – a very large urban area that forms when large urban areas merge

mid-latitudes – the two bands of Earth that lie in latitudes 30° – 60° north of the Equator and 30° – 60 south of the Equator

migrant – a person who moves to another part of the country, or another country, usually to work

Milankovitch cycles – changes in Earth's movements, which affect the climate

millibar – air pressure is measured in this unit

monsoon rains – rains that fall in summer in some regions, when moist winds are drawn in from over the ocean

N

natural – occurs without human involvement

natural increase – the number of births minus the number of deaths in a period (eg a year)

non-renewable resource – a resource we will run out of one day; for example oil

North Atlantic Drift – a warm current in the Atlantic Ocean; it keeps the weather on the west coast of Britain mild in winter

O

ocean currents – currents of water in the ocean that are warmer or colder than the water around them

one-party state – one political party governs the country; other parties may be banned

P

peninsula – land that juts out into the sea, and is almost surrounded by water

physical geography – deals with Earth's natural features and processes; for example coasts, glaciers, and climate

plateau – an area of fairly flat high land

population – how many people live in a place

population density – the average number of people living in a place, per square kilometre

population distribution – how the people in a country are spread around

population pyramid – a diagram showing the population in different age groups

populous – has a large population

postcode – a set of numbers and letters which are added to an address to help mail delivery

precipitation – water falling from the sky (as rain, sleet, hail, snow)

prevailing wind – the wind that blows most often; in the UK it is a south west wind (blowing *from* the south west)

pull factors – factors that attract people to a place (for example, better jobs)

push factors – factors that push people out of a place (for example, there's no work there)

Q

Quaternary period – the period from 2.6 million years ago till today

R

rainforest – has lush vegetation, with many different species of plants and animals

rate of infiltration – how fast water soaks into the soil

regenerate – to restore an area that was in a poor state, and bring it back to life

relief – how the height of the land varies

relief rainfall – forms when moist winds are forced to rise, when they meet high land

renewable resource – a resource we won't run out of; for example sunlight, wind, rivers

renewables – term used for renewable sources of energy (such as sunlight, for solar power)

republic – a country that has no king, queen, emperor, or other type of royal family

resources – things we need to live, or use to earn a living; for example food, fuel

rural area – an area that is mainly countryside; it may have villages and small towns

S

salt marsh – a low-lying marshy area by the sea, with salty water from the tides

settlement – a place where people live; it could be a hamlet, village, town or city

shingle – small pebbles

slum – area of very poor housing

social – about people and society

solar power – when we use sunlight to generate electricity (via solar panels)

sparsely populated – few people live there

species – the group a plant or animal belongs to

speculators – they take a risk, and spend money in the hope of making lots of profit

spit – a strip of sand or shingle in the sea

squatter settlement – a slum made up of shacks that people built illegally

stack – a pillar of rock left standing in the sea when the top of an arch collapses

steppe – a large flat area of treeless grassland

storm surge – an abnormal rise in sea level caused by a storm; it brings coastal floods

stump – the remains of an eroded stack

sustainable – can be carried on into the future without harming people's quality of life, or the economy, or the environment

T

taiga – region of coniferous forests which lies between the tundra and steppes

temperate – describes a mild climate: not too hot, not too cold

tides – the natural rise and fall in sea level, due mainly to the pull of the moon

transport – the carrying away of material by rivers, waves, the wind or glaciers

tree line – the height or altitude above which it's too cold for trees to grow

tropical cyclone – another name for a hurricane

tropics – the area between the Tropics of Cancer and Capricorn

troposphere – the lowest layer of the atmosphere

tundra – a cold region where the ground is deeply frozen; only the surface thaws in summer, allowing small plants to grow

typhoon – another name for a hurricane

U

urban area – a built-up area (a large town or a city); it's the opposite of rural

urbanisation – the increase in the % of the population living in urban areas

W

water vapour – water in gas form

wave-cut notch – a notch cut in a cliff face by the action of waves

wave-cut platform – the flat rocky area left behind when waves erode a cliff away

weather – the state of the atmosphere at any given time; for example how warm it is

weathering – the breaking down of rock; it is caused mainly by the weather

wind – air in motion

wind direction – where the wind blows *from*

windward – facing into the wind

Index

A
air masses 76
air pressure 74
altitude 85
arch 56
artificial reef 66
Asia 105–118
atmosphere 71

B
backwash 52
bay 56
beach 57
beach nourishment 66
Belt and Road Initiative 134
biomes (of Asia) 116–117
birth rate 27

C
carbon dioxide 98
cave 56
China 119–136
cholera 6
cliffs 56
climate 84–89
climate change 91–104
climate graph 87
climate regions (world) 88
clouds 81
coasts 51–68
coastal defences 66
coastal landforms 56–57
condenses 80
continent 106
Coriolis effect 72
crime (and GIS) 16–17

D
dams 125
data 7
death rate 27
decline (industry) 36
deindustrialisation 36
deposition (by the waves) 55
depression (weather) 78–79
desert (cold and hot) 88
dew 75
distance from sea (climate) 85
drought 97

E
emigrants 100–101
emissions 100–101
enquiry question 8
erosion (by the waves) 54
Ethiopia 31
ethnic groups (Asia) 110

F
fertility rate 24
fetch (wind) 52

fieldwork 8–13
fog 75
fronts (cold and warm) 77
frost 75

G
GIS (Geographic Information System) 14–15
global atmospheric circulation 72
global warming 96
GNI per person (PPP) 24
Green Great Wall 133
greenhouse gases 98
groynes 66

H
Happisburgh, Norfolk 64–65
headland 56
high pressure weather 74–75
Hong Kong 120
hurricane 82
hypothesis 8

I
ice ages 92
immigrants 26
industrialisation 36
Industrial Revolution 34

J
Japan 31

L
latitude 84
life expectancy 20
longitude 14
longshore drift 55
low pressure weather 74

M
Macau 120
Manchester 38–41
mangrove swamps 117
megalopolis 129
methane 98
mid-latitudes 76
Milankovitch cycles 95
mountain (biome) 117

N
natural increase 26
Newquay, Cornwall 60–61

O
ocean currents 73
ocean sediment 94
one-child policy (China) 127

P
Plateau of Tibet 124
population 19–32
population density 114

population distribution 22
population pyramid 27
prevailing winds 72
pull factor 43
push factor 43

Q
Quaternary period 92

R
rainfall 80–81
regeneration (urban area) 38
resources 28
revetments 66
rock armour 66
rural area 35

S
salt marsh 57
sea walls 66
seasons 84
settlements 34
Shenzhen 128–129
slums 45
spit 57
special economic zones 128
squatter settlements 45
stack 56
steppes 116
storm surge 62–63
stump 56
sustainable 48
swash 52

T
taiga 116
Taiwan 120
temperate forest 116
Thames Barrier 67
tides 53
thunderstorms 75
transport (by the waves) 55
tree rings 94
tropical cyclone 82
troposphere 71
tundra 116

U
uprush 52
urban areas 35
urbanisation 35
urban sprawl 37

W
warm moist forest 117
water vapour 71
wave-cut notch 56
wave-cut platform 56
waves 52
weather 70–83
wind 71